THE 50 GREATEST ENGINEERS

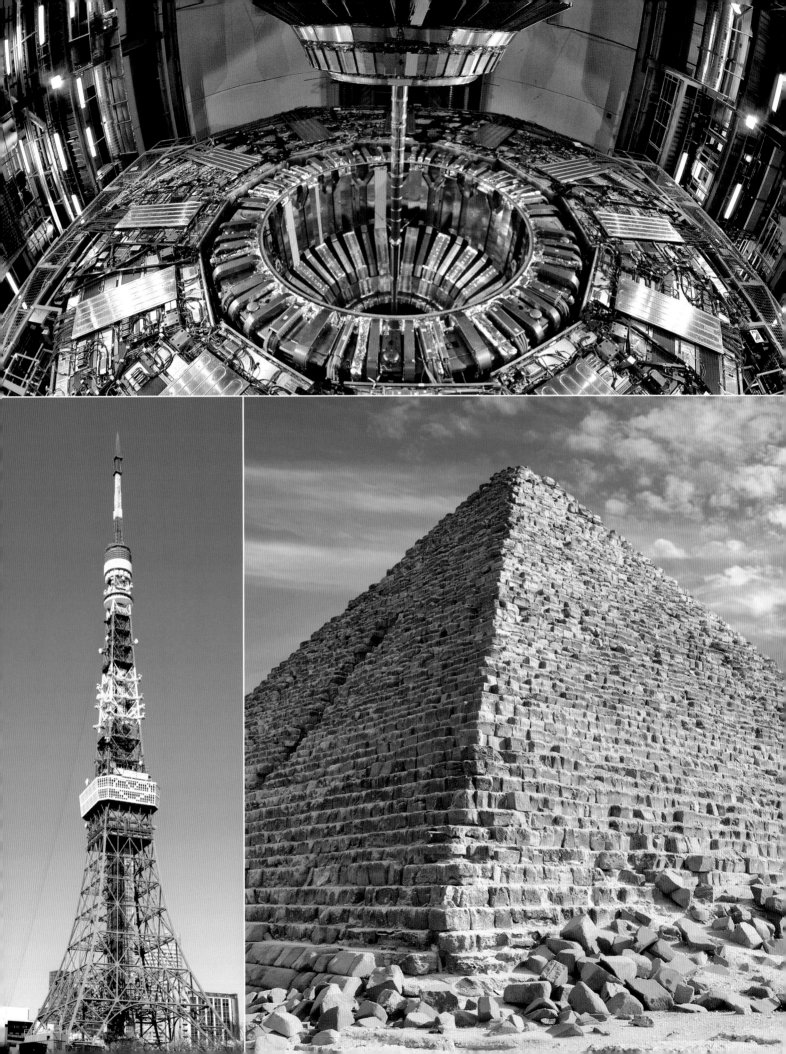

THE 50 GREATEST ENGINEERS

THE PEOPLE WHOSE INNOVATIONS HAVE SHAPED OUR WORLD

PAUL VIRR AND WILLIAM POTTER

ARCTURUS

ARCTURUS

This edition published in 2021 by Arcturus Publishing Limited
26/27 Bickels Yard, 151–153 Bermondsey Street,
London SE1 3HA

ISBN: 978-1-83857-421-5
AD008039UK

Printed in China

CONTENTS

INTRODUCTION

When you hear the word 'engineering', perhaps the first images that spring to mind are feats of construction, such as iconic bridges, buildings or vehicles. But beyond these showcase achievements lies a more general truth: engineering is everywhere in the human world. From towering skyscrapers and the Large Hadron Collider (a vast scientific instrument that is the largest machine in the world), right down to the invisibly small circuits on silicon chips and microscopic carbon nanotubes, our world, the human world that we inhabit, has all been engineered. This book tells the stories of just a selection of engineers who have contributed to the construction of this world.

Before we meet some of these engineers, it is worth examining the work that they do, even to ask the basic question: what is engineering? Engineering is a word often preceded by a qualifying term, such as civil, structural or mechanical, just to name a few. This array of specialisms reveals how wide the field is. However, one thing that all engineers have in common is that they provide practical solutions to real-world problems. Their solutions are often physical structures and machines, but can also be less tangible, such as processes to turn raw materials into useful products, or to transform data into meaningful information.

To unpack the idea of engineering further, we can say that engineers supply the means to meet all kinds of human needs. While evolution has shaped all living things to suit the conditions they live within, we humans have a unique set of skills that have allowed us a greater degree of control over the natural environment. We have used our intelligence to make tools and with these we have shaped the world to

BELOW: The Large Hadron Collider at CERN, a marvel of modern engineering.

ABOVE: Filippo Brunelleschi's dome for Santa Maria del Fiore was one of the crowning feats of Renaissance engineering.

suit us. Perhaps we can consider our prehistoric ancestors, who fashioned stone hand-axes from flint to be the first engineers. From their first, basic tools through to simple machines, such as levers, pulleys and the wheel, we can trace the foundations of the engineering that has finally led us to the modern world where humans are the dominant species on the planet. The everyday infrastructure that surrounds us today is the cumulative result of engineering history. If you stop to consider any of the things you use every day, you are sure to find it has a fascinating engineering back story.

Not all engineers are pioneers, nor are they always scientists and inventors, though history has plenty of examples of engineers who have encompassed all these roles. Engineers often build on the work of others, combine or refine existing inventions to deliver an engineered product. Hand-in-hand with scientific discoveries, engineers have applied science

to the practicalities of everyday life. This is why the engineering story has been particularly rich and varied during periods of intense scientific advances, such as the Renaissance and the Industrial Revolution.

We have chosen 50 engineers that represent some of the key engineering achievements from ancient to modern times. With a subject as vast as engineering we have had to be necessarily selective. It is only relatively recently that more of the achievements of women engineers are being revealed from the archives. It is to be hoped that their successes and those of all the other engineers in this book will prove inspirational. Now, perhaps more than ever, humankind needs engineers to face the challenges of the future, particularly the effects of climate change. They may deploy new skills, such as robotics and artificial intelligence, but they will be finding solutions to problems and making their ideas reality, just the same as all the engineers you will meet here.

IMHOTEP

'FROM THE HEIGHTS OF THOSE PYRAMIDS, FORTY CENTURIES LOOK DOWN UPON US.'

Napoleon Bonaparte, 1798

GREATEST ACHIEVEMENTS

PYRAMID OF DJOSER, SAQQARA
Completed c. 2650 BCE, it was the first step pyramid and one of the earliest examples of monumental architecture.

ABOVE: A statuette of Imhotep from the 7th century BCE.

Of the Seven Wonders of the Ancient World, only the Great Pyramid of Khufu is still standing today. It is the largest of more than one hundred pyramids that were built as tombs for the pharaohs who ruled ancient Egypt. Exactly how these colossal funeral monuments were built is still the subject of lively debate among Egyptologists, engineers and experimental archaeologists. But one thing the experts all agree on is that these ancient structures are a testament to the skills of the early engineers that built them.

The pyramids and temples of ancient Egypt have immortalized the names of the rulers who commissioned them, but they have also ensured that the names of at least some of the pyramid builders have not been forgotten. A handful of statues and inscriptions have survived and they give us a tantalizing glimpse of some of these early civil engineers – the first in recorded history. Foremost among the early pyramid builders was Imhotep. He was the vizier for Djoser, one of the first kings of the Third Dynasty. As the pharaoh's most senior official, Imhotep was responsible for the day-to-day running of the kingdom, but he also managed all the royal building works. It was in this role that Imhotep built the first pyramid, more than 4,600 years ago, at Saqqara.

Unlike the later pyramids at Giza, which have flat sides rising from a squarish base to the apex, this very first pyramid was a stepped structure made from tiers of progressively smaller platforms. It stood at the centre of a huge funerary complex commissioned by

BELOW: Despite centuries of wear, the stepped structure of the Pyramid of Djoser at Saqqara remains clearly visible today.

ABOVE: The Hypostyle Hall built by Imhotep as part of the Saqqara necropolis surrounding Djoser's pyramid.

Djoser, which was just part of an ambitious building programme that extended throughout his kingdom. As the location of ceremonies that celebrated his coronation, Djoser's tomb was the pinnacle of a series of architectural statements that reflected and cemented his living power.

The Saqqara Step Pyramid that Imhotep designed and built was a revolutionary structure – the world's first large building built from stone. Prior to this, all the buildings in ancient Egypt had been built from mud-brick, reeds and wood. The pyramid also marked a radical change in the way the funeral monuments of ancient Egypt's rulers were built. The tombs of earlier pharaohs were flat-roofed, rectangular buildings called mastabas. They stood about 9 m (30 ft) high with flat roofs and sloping walls made of mud-brick. Imhotep broke away from tradition and built a royal tomb from limestone blocks on a much more imposing scale.

Utilizing more than 300,000 m³ (10 million ft³) of limestone, this was the largest and most complicated civil engineering project that Egypt had seen to date. Nothing like the Step Pyramid had been built before, so it became the prototype that was developed by later pyramid builders. Alongside the technical challenges, Imhotep had to deal with the logistics of sourcing and transporting materials, and marshalling a large workforce. Contrary to classical accounts, the labour force that built Egypt's pyramids was not made up of slaves. It is likely that Imhotep used a semi-permanent core of skilled workers, supplemented by a rotating workforce over the years.

Djoser's pyramid was essentially six limestone mastabas, each smaller than the one below, stacked on top of each other. It was built in stages, starting with a traditional mastaba that was then encased in limestone. Subsequent platforms of limestone blocks were then built on this foundation. When completed, the pyramid stood 62.5 m (205 ft) above its rectangular base and was a towering presence that could be seen

right across the Saqqara plateau. Around the pyramid was a series of temples and buildings devoted to rituals. This huge complex was surrounded by a limestone wall more than 10 m (33 ft) high. The engineering works below ground were similarly impressive. A labyrinth of tunnels surrounded the king's granite burial chamber and there were hundreds of rooms, some with ornately decorated columns like bundled papyrus reeds or lined with blue tiles.

It took Imhotep about 18 years to complete the Step Pyramid. This monumental engineering work was achieved without wheels or pulleys, using levers and simple tools such as copper chisels, saws and drills, rounded hammer stones, plumb lines and measuring sticks. Imhotep's landmark structure was an inspiration to the pyramid builders that followed him and still inspires engineers today.

LEFT: A statue of the pharaoh Djoser.

Completed around 2560 BCE, the Great Pyramid at Giza was a huge tomb for the pharaoh Khufu. Built from about 2.3 million stone blocks, it was more than 140 m (460 ft) high on completion and remained the tallest human-built structure on the planet for more than 3,800 years. Khufu's vizier, Hemiunu, was the chief engineer and architect responsible for the construction of the Great Pyramid.

RIGHT: Khufu's vizier, Hemiunu, was chief engineer and architect of the Great Pyramid.

BELOW: The Great Pyramid of Khufu near Cairo.

ARCHIMEDES

'GIVE ME A PLACE TO STAND AND I WILL MOVE THE EARTH.'

Archimedes

GREATEST ACHIEVEMENTS

ARCHIMEDES' PRINCIPLE
MID-3RD CENTURY BCE

COMPOUND PULLEY c. 250 BCE

ARCHIMEDES SCREW c. 250 BCE

THE CLAW OF ARCHIMEDES
Designed in 214 BCE by Archimedes during the Siege of Syracuse to protect the city from Roman invaders.

ABOVE: Archimedes.

Although much is uncertain about the life and work of the ancient Greek mathematician Archimedes, thanks to classical accounts, some written centuries after his death, he is remembered not only as a trailblazing mathematician, but also as one of the first engineers in history.

Archimedes was born in the 3rd century BCE at Syracuse, a Greek city-state on the coast of what is the island of Sicily today. By this time the central influence of Athens was waning, but Greek intellectual culture had spread and was thriving all around the Mediterranean. As a curious young man, eager for knowledge, Archimedes probably travelled to Egypt for his education.

At that time the Egyptian city of Alexandria was the intellectual hub of the Mediterranean and home to the Mouseion, a renowned centre of learning and research. At its heart was the Great Library of Alexandria, which held tens of thousands of papyrus scrolls. This treasure-house of knowledge attracted scholars from far and wide. We know Archimedes corresponded with its chief librarian, the Greek astronomer Eratosthenes, but can only guess at the other intellectual influences he came into contact with. Archimedes would have met scholars who shared his passion for mathematics and geometry, but also engineers developing new military technology for their rulers.

When Archimedes returned home, he divided his time between mathematics and work as an engineer in the service of Hiero II, ruler of Syracuse. Archimedes' mathematical achievements are preserved in his writings and are largely theoretical. His practical

BELOW: Water being raised with an Archimedes screw.

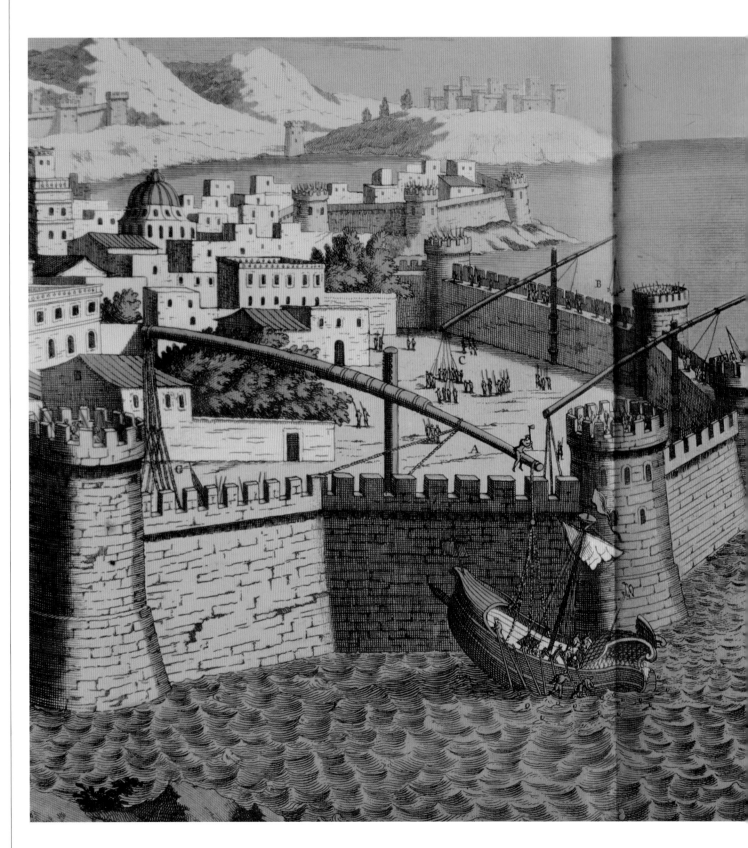

talents as an inventor and engineer of mechanical genius are revealed in a series of stories.

One of the most famous of these stories relates how King Hiero II tasked Archimedes with working out if a crown made for him was pure gold, or if the goldsmith had stolen some of the gold and substituted an equal weight of less dense silver. This would have meant the crown weighed the same, but occupied a

LEFT: The Claw of Archimedes in action against Roman invaders at the siege of Syracuse.

idea while lowering himself into a bath and seeing the water level rise. Sadly, the famous story about him running naked into the street shouting 'Eureka!' ('I've found it!'), was probably a fanciful embellishment.

Other stories credit Archimedes with engineering devices that solved a variety of real-world challenges. Archimedes is said to have invented the compound pulley, which he publicly demonstrated by hauling a ship onto the shore single-handed. Archimedes may also have used pulleys for the world's first elevator. He is also famous for creating the Archimedes screw, a type of pump with a helical screw housed inside a cylinder. Both of these were revolutionary in the ancient world and are still two of the simple machines that underlie many mechanical devices today.

Archimedes is also closely associated with the lever, which he probably saw being used in construction projects in Egypt. Although Archimedes didn't invent the lever, he did describe how they worked mathematically. He also famously claimed that with a long enough lever and a pivot he could move the world.

Like many engineers throughout history, Archimedes was called upon to use his talents in the service of war. In order to defend his home city of Syracuse, Archimedes improved the power and accuracy of its catapults. He is also said to have built poles that extended from the city walls to drop boulders on enemy ships. He may also have built a device called the Claw of Archimedes that could snatch ships out of the water and shake them to pieces. An even more fantastical weapon he is said to have invented was a heat ray that used mirrors to focus the sun's energy on enemy vessels and set them alight.

Unfortunately, Archimedes' military inventions could not save Syracuse from the Romans. The city fell after a two-year siege and Archimedes' incredible ingenuity was brought to an end when a Roman soldier killed him. The diverse achievements attributed to Archimedes may blur mythology and history, but they reveal a thinker who solved real-world problems through engineering.

different volume. Archimedes realized that he could work out the volume occupied by an irregular object by using it to displace an equal volume of water from a container. He is reported to have come up with this

CTESIBIUS

'I MUST TELL ABOUT THE MACHINE OF CTESIBIUS, WHICH RAISES WATER TO A HEIGHT.'

Vitruvius, *De Architectura*

ABOVE: Ctesibius.

Just over 2,300 years ago, Alexandria in Egypt was a global centre of learning. Scholarship, technology and engineering thrived under the rule of the Ptolemies, a dynasty of Greek origin, descended from one of Alexander the Great's generals. They founded a great library to collect all human knowledge, as well as a great centre for research and teaching called the Mouseion. Its first director was Ctesibius, a Greek inventor and mathematician who became one of the founding figures of engineering in the ancient world.

Ctesibius lived in Alexandria during the 3rd century BCE. Like other scholars in the city, he probably had the support of Ptolemy II as a wealthy patron: many of the surviving manuscripts from this era bear dedications to the presiding ruler. Sadly, the writings of Ctesibius perished in antiquity, but we know of his work through the writings of later scholars, such as Heron of Alexandria and Archimedes, who cite him in their texts. Evidence suggests that Ctesibius wrote an influential treatise on pneumatics, describing the elasticity of air and how compressed air could be used in practical devices, such as pumps. He also wrote about the science of hydrostatics, describing the

mechanics of fluids and pointing towards its practical applications. The invention of the siphon has been attributed to Ctesibius by some classical writers.

Ctesibius is thought to have been the son of a barber. One story describes how he invented a height-adjustable mirror for his father's shop using pulleys and a lead counterweight enclosed in a cylindrical pipe. Apparently the movement of this weight

RIGHT: Ctesibius' clepsydra, an ingenious water-powered clock.

RIGHT: Ctesibius' force pump could raise water from wells or produce a jet of water for irrigation, fountains or even to put out fires.

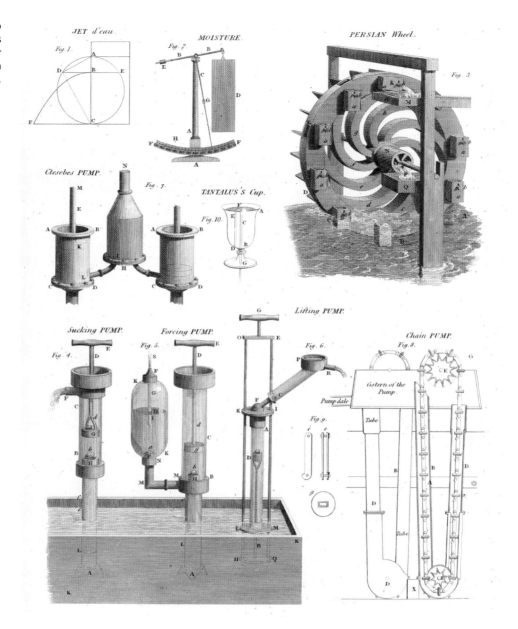

compressed the air in the tube and created a musical tone, inspiring the young Ctesibius with ideas for some of his later engineering innovations. One of the most useful was his invention of a force pump that used pistons, cylinders and valves to raise water. Described as 'Ctesibius' machine' by the Roman architect Vitruvius in *De Architectura*, we have archaeological evidence for Roman force pumps based on Ctesibius' design. These were the only pumps from antiquity capable of producing a jet of water under pressure. As such, they were useful for fountains, but could have been used to douse fires.

Ctesibius also designed the hydraulis, a predecessor of the modern pipe organ. It used water pressure to drive compressed air through pipes of various lengths, making sounds of correspondingly different pitches. Another of Ctesibius' devices that relied on water for its operation was the clepsydra – a type of clock that used dripping water to gradually fill a container below, raising a float device that pointed to the hours. Ctesibius improved the clepsydra to maintain a steady flow of water, which made it the most accurate timepiece on the planet until the invention of the pendulum clock in the 17th century. Literally ahead of his time, Ctesibius' pioneering theoretical and practical innovations led the way for the generations of engineers that would follow.

VITRUVIUS

'THERE ARE THREE DEPARTMENTS OF ARCHITECTURE: THE ART OF BUILDING, THE MAKING OF TIMEPIECES AND THE CONSTRUCTION OF MACHINERY.'

Vitruvius, Book I, *De Architectura*

GREATEST ACHIEVEMENTS

DE ARCHITECTURA
c. 30–15 BCE
Vitruvius' manuscript was a wide-ranging manual on Roman engineering that inspired architects for centuries to come.

BASILICA AT FANO *c.* 19 BCE
Chief Engineer and Architect for Julius Caesar and the Emperor Augustus late 1st century BCE

ABOVE: Vitruvius presenting *De Architectura* to Augustus.

RIGHT: Reconstruction of a Roman ballista, as used at the Siege of Alesia.

In the 1st century BCE, during the reign of the emperor Augustus, a former Roman soldier and architect named Vitruvius retired from service. Vitruvius had been generously granted a pension, possibly thanks to the patronage of the emperor's sister Octavian, and he might well have languished in obscurity or have become a minor footnote in Roman history, but for one thing: Vitruvius decided to write a book.

With an eye on posterity and intent on preserving the knowledge he had acquired during a long career as an architect, military and civil engineer, Vitruvius devoted his retirement to writing a treatise on architecture. Entitled *De Architectura*, its main subject was architecture, but as this field was much broader in antiquity than today, it embraced a wide range of engineering topics and artisanal practices. For this reason, *De Architectura* provides us with an unparalleled insight into Roman engineering as Vitruvius discusses everything from building temples, theatres and aqueducts, through to clocks, siege weapons and water pipes.

Vitruvius' treatise eventually ran to ten books and is the only substantial account of classical

ABOVE: Julius Caesar laying siege to the fortified settlement of Alesia, Gaul (a region now within modern-day France and Belgium).

architecture to have survived to the modern day. It was much-copied in its time, though some of these versions were incomplete or of an inferior quality. Vitruvius' text was originally illustrated, but these important accompaniments have sadly been lost through the ages. Some of his later readers have found Vitruvius' technical descriptions wanting in clarity, but nevertheless *De Architectura* has proved a technical resource for generations of engineers and architects.

De Architectura went on to become the most influential book on architecture from antiquity. This is largely due to the Italian scholar Poggio Bracciolini, who rediscovered a complete copy of *De Architectura* in 1414 in the library of a Swiss abbey. Bracciolini incorporated some of Vitruvius' text into his own work. Translations of the original followed. Through these versions Vitruvius became an inspiration to architects and engineers during the Renaissance, including Leonardo da Vinci. Vitruvius' idea that a structure should exhibit the three qualities of stability,

utility and beauty, was particularly influential during the Renaissance. Known as the Vitruvian Triad, it still informs modern-day architectural practice.

Details of Vitruvius' life and career are limited. What we know has mostly been gleaned from the text of his book. Vitruvius seems to have started out as a military engineer serving under Julius Caesar during the Gallic Wars. He would have been responsible for catapults, ballistae that fired projectiles and various machines for siege warfare, such as battering rams and siege towers for mounting attacks on fortresses. In his memoirs, Caesar quotes Vercingetorix, the defeated leader of the Gauls, who ascribed the Romans' victory to their skills in siege warfare, underlining the importance of military engineers. Regarding Vitruvius' later career as an architect and civil engineer, we only know of a basilica he built at Fano, but that no longer exists. However, in place of physical monuments to his engineering genius, we have his book, which continues to be read after more than 2,000 years.

HERON OF ALEXANDRIA

'PLACE A CAULDRON OVER A FIRE: A BALL SHALL REVOLVE ON A PIVOT.'

Description of the first steam engine, from Heron of Alexandria's *Pneumatics*

GREATEST ACHIEVEMENTS

DIOPTRA 1ST CENTURY CE
A forerunner of the modern theodolite.

ODOMETER 1ST CENTURY CE
A measuring device to record distance travelled.

MECHANICA 1ST CENTURY CE
A book containing information on many machines, such as levers, pulleys, gears and cranes.

PNEUMATICA 1ST CENTURY CE
Includes descriptions of automatons and other mechanical devices.

AEOLIPILE 1ST CENTURY CE
The world's first working steam engine.

ABOVE: Heron of Alexandria demonstrates the aeolipile.

The city of Alexandria in Egypt had been established as a centre of learning in the 3rd century BCE by its Greek rulers, the Ptolemies. They founded Alexandria's legendary library, designed to collect all human knowledge, as part of a complex devoted to learning called the Mouseion. Heron of Alexandria, one of the greatest engineers of antiquity, worked here in the 1st century CE when the city was under Roman rule.

Little is known of Heron (or Hero) of Alexandria. Born in the early part of the 1st century, he was either Greek or an Egyptian who had received a Greek education. Heron is thought to have taught at the Mouseion because his surviving works are written as lecture notes. Scholars at the Mouseion drew on the knowledge that flowed into Alexandria from ancient Egypt, Babylonia, Rome and Greece. Heron's work builds on that of his predecessors, including Archimedes and Ctesibius. Just a handful of Heron's writings survive, but they reveal him as one of the figures of early engineering. His achievements range from applied mathematics to mechanical and civil engineering.

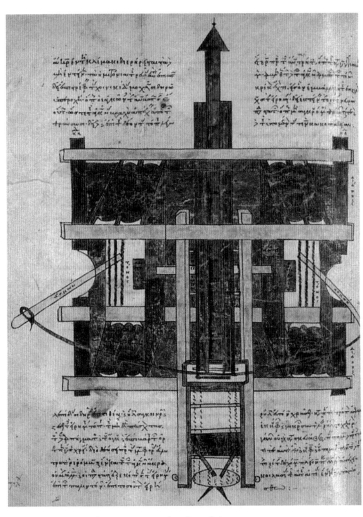

ABOVE: Illustration from a manuscript describing Heron's design for a cannon.

As a mathematician, Heron's work included geometry and the practical applications of maths in measurement, land surveying and civil engineering. In his work *On the Dioptra*, Heron presents some key tools for civil engineers and surveyors, including an adaptation of an optical surveying instrument called a dioptra. This was used by the ancient Greeks to make astronomical measurements, but Heron's improved version could also be used for more down-to-earth applications, such as land surveying. A forerunner of the modern theodolite, Heron's dioptra could measure angles, lengths and heights from a distance. It was useful for finding the depth of a ditch, the width of a river, or for planning how to tunnel through a hill from both sides so as to meet in the middle. He also describes a mechanical measuring device called an odometer. This specially adapted carriage could quickly measure horizontal distances. It had standard-sized Roman chariot wheels and used gears connected to the axle to turn pointers on a series of dials. These quickly and accurately recorded the distance travelled.

In another work, *Mechanica*, Heron discusses methods for lifting heavy weights. These include an array of simple machines, including levers, pulleys, wedges and gears, as well as more sophisticated machinery such as cranes. *Mechanica* is almost a technical manual for civil engineers and complements Heron's writings on land surveying tools in *On the Dioptra*. His work would have been invaluable to Roman civil engineers constructing aqueducts, tunnels and buildings throughout the empire.

Heron's talents as a mechanical engineer are revealed in his *Pneumatica*. This is a compendium of mechanical devices driven by air, steam or water pressure. Many of his machines are automata. Some were devised as spectacles for temples, such as automatically opening doors. Others were created for entertainment, including a fully-automated theatre. Devices like these, which appeared to work independently, would have amazed their audiences. They also prefigured later developments in robotics. One of Heron's inventions was in fact an early robot – an automated cart used for

theatrical entertainment. It was powered by a falling weight attached to a string wrapped around its two axles. As the weight fell, the string unwound and turned the wheels. A series of pegs could be nailed into the axles to change the direction that the wheels turned. This made Heron's cart the first programmable device that we have on record. His mechanical robot could be programmed to start, stop and turn, making Heron a pioneering computer engineer, too!

Amid Heron's numerous engineering achievements, the device he is most famous for is the aeolipile. Described in his *Pneumatica*, this was the world's first detailed account of how to build a working steam engine. The aeolipile was a hollow metal sphere, mounted like a horizontal spindle so that it could freely rotate. The sphere was fed with steam from a boiler below it. This steam then shot out of the sphere under pressure through two curved nozzles that faced in opposite directions. This provided the power to make the sphere rotate. Heron's aeolipile operated on the same principle as the steam engines that would power the Industrial Revolution more than a thousand years later.

RIGHT: A reconstruction of Heron's odometer, a mechanical distance-calculating carriage.

ZHANG
HENG

'[HIS] MATHEMATICAL COMPUTATIONS EXHAUSTED THE RIDDLES OF THE HEAVENS AND THE EARTH. HIS INVENTIONS WERE COMPARABLE EVEN TO THOSE OF THE AUTHOR OF CHANGE. THE EXCELLENCE OF HIS TALENT AND THE SPLENDOUR OF HIS ART WERE ONE WITH THOSE OF THE GODS.'

Memorial inscription written by Zhang Heng's friend, Cui Ziyu

GREATEST ACHIEVEMENTS

WATER CLOCK EARLY 2ND CENTURY CE

SOUTH-POINTING CHARIOT EARLY 2ND CENTURY CE

ODOMETER c. 125 CE

ARMILLARY SPHERE c. 125 CE

ESTIMATE OF THE VALUE OF PI c. 130 CE

SEISMOSCOPE c. 132 CE

ABOVE: Zhang Heng.

Like many scholars from antiquity, the Chinese court official, astronomer and mathematician Zhang Heng's many talents included engineering. In the ancient world, long before engineering had become a distinct profession or a separate discipline, it was part of a spectrum of skills and knowledge that scholars put at the disposal of their rulers and patrons.

Zhang Heng's accomplishments ranged between the philosophical and practical. He was a renowned poet and widely-read author. Among his mathematical achievements, Zhang Heng came up with a working estimate for the value of pi (π). As a mechanical engineer and inventor, Zhang Heng used hydraulics and gears to build innovative scientific instruments to measure time and predict astronomical movements. He is also credited with inventing the world's first seismoscope. His device was not only capable of

registering when a distant earthquake occurred, but could also indicate the direction from where the geological disturbance originated.

Zhang Heng was born in 78 CE, in Nanyang, a city in the south-west of Henan Province, central China. He came from a reasonably wealthy and influential family, with links to the civil service and the court. The young Zhang Heng grew up and received his education during the 1st century, when the Eastern Han Dynasty ruled. It was a peaceful and prosperous period, considered to be a golden age. Culture and ideas flourished with the political stability of Emperor Zhao's rule.

The young Zhang Heng was sent to study in Chang'an (modern-day Xi'an) and then Luoyang, the capital of the Eastern Han Dynasty. This was an established educational centre for the families of civil servants. Luoyang was at the eastern end of the Silk Road, an international trade route that made the city a cultural as well as a commercial hub. Traders brought new ideas with them, as well as goods, so Zhang Heng was well-placed to experience the latest thinking and technology.

Following his studies, Zhang Heng returned home at the age of 23 to take up an administrative position in local government. He also continued to write poetry and furthered his studies in mathematics and astronomy. News of Zhang Heng's abilities filtered through to Emperor An, the sixth Han Emperor. He summoned Zhang Heng, now in his early thirties, to the court, where he eventually became the Emperor's Chief Astronomer. Disagreements held Zhang Heng's career back, but having fallen out of favour, he bounced back and was reappointed Chief Astronomer by the seventh Han Emperor, Emperor Shun.

Astronomy was used to regulate the calendar and predict auspicious and inauspicious days in ancient China, and so it played a fundamental part in scheduling and decision-making. Zhang Heng made key improvements to two inventions that helped with imperial time-keeping: a water clock known as a clepsydra and the armillary sphere, which was used to predict the movements of planets and stars in the night sky.

A clepsydra worked by water slowly dripping from a reservoir into a collecting vessel below, which had a floating indicator whose pointer told the time as it rose. However, as the reservoir was depleted, the water pressure decreased, slowing the flow and making the clock run slower. To solve this problem, Zhang Heng ingeniously kept the flow steady by using an extra compensating tank below the reservoir.

Zhang Heng also made key contributions to the armillary sphere. Firstly, he added two additional rings to improve how it was calibrated. Secondly, he used his knowledge of hydraulic engineering to automate the

LEFT: An ornately decorated armillary sphere in the courtyard of the Beijing Ancient Observatory, China.

LEFT: Ancient satnav: a model of Zhang Heng's south-pointing carriage.

such, they reflected on the emperor, so it was important for him to be informed about even the most-distant earthquakes. Zhang Heng's seismoscope consisted of a heavy brass container with eight dragons mounted on the outside. These were surrounded by eight corresponding toads on the base. Each dragon had a brass ball, finely balanced in its mouth. Tremors from an earthquake triggered the mechanism inside the central vessel, causing a ball to drop from the mouth of one of the dragons into the mouth of the toad below it. This signalled that an earthquake had happened and indicated the direction from which it originated.

Zhang Heng was nearing the end of his career when he demonstrated his device to the emperor, who was so pleased that he more than tripled Zhang Heng's salary. Zhang Heng died seven years later at the age of 61, a much-respected figure whose mechanical innovations inspired later engineers. After his death, the poet Fu Xuan paid tribute to Zhang Heng, saying it was a shame that his engineering talents hadn't been more widely used.

BELOW: Ancient earthquake detector: a cutaway model of Zhang Heng's seismoscope showing its internal seismoscope.

armillary sphere so that it could mimic the movements of objects visible in the night sky. To power the clever gearing mechanism of his automated armillary sphere, Zhang Heng used a water wheel. He may have been inspired to do this by the iron foundries of his hometown Nanyang. These had used water wheels to pump the huge bellows that blew air into the furnaces. Zhang Heng's design innovations would lead to further developments and the building of large clock-tower-style armillary spheres by subsequent engineers, such as Su Song in the 11th century.

Sophisticated gearing was the basis of another invention attributed to Zhang Heng: a south-pointing chariot. This was a kind of mechanical compass mounted on a chariot. It used differential gears driven by the wheels to rotate a figure with an outstretched hand that always pointed south. This navigational device was a kind of ancient satnav, and although limited it would have been invaluable for directing troop movements and land surveying. Another chariot-based device, the odometer, which was used to measure distances, has also been credited to Zhang Heng. Gears powered by the chariot's wheels moved automated figures that struck a drum roughly every half km (0.3 miles) and a gong every 5 km (3 miles).

Possibly the crowning achievement of Zhang Heng's engineering career was his invention of the world's first seismoscope. Earthquakes were very important in ancient China as they were seen as a sign of divine displeasure with activities in the human world. As

APOLLODORUS OF
DAMASCUS

ABOVE: Apollodorus of Damascus.

GREATEST ACHIEVEMENTS

TRAJAN'S BRIDGE 105 CE

TRAJAN'S FORUM 112 CE

TRAJAN'S COLUMN 113 CE

ABOVE: Apollodorus of Damascus.

'THE ROMAN EMPEROR TRAJAN … SEEMED TO BE FILLED WITH RESENTMENT THAT HIS REALM WAS NOT UNLIMITED, BUT WAS BOUNDED BY THE DANUBE RIVER. HE WAS EAGER TO SPAN IT WITH A BRIDGE SO THAT HE MIGHT CROSS IT AND THAT THERE WOULD BE NO OBSTACLE TO HIS BATTLING THE BARBARIANS BEYOND IT. HOW HE BUILT THIS BRIDGE I SHALL NOT BE AT PAINS TO RELATE, BUT SHALL LET APOLLODORUS OF DAMASCUS, WHO WAS THE MASTER-BUILDER OF THE WHOLE WORK, DESCRIBE THE OPERATION.'

Procopius, Byzantine Greek historian, 6th century

BELOW: Illustration of an armoured siege engine designed by Apollodorus.

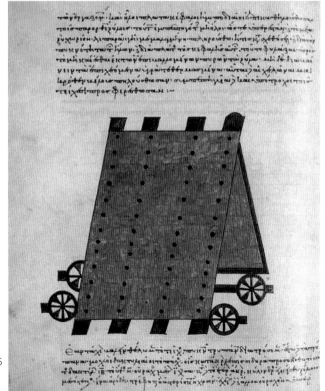

Little is known about the engineers who built the great cities and monuments of the ancient world. Some of the structures they designed have survived and still stand today. We can stand and admire these engineering marvels, but sadly, we only know the identity of a handful of the engineers who built them. Fortunately, something of the career of Apollodorus of Damascus has been preserved in historical accounts. Apollodorus rose to prominence as the engineer of large-scale public projects for the emperor Trajan, including the building of Trajan's Forum in the heart of Rome. He is also remembered for his early career as a military engineer. Apollodorus is also thought to be the author of a treatise entitled *Poliorketika*, an engineering manual that describes how to build and use various siege weapons.

ABOVE: Relief from Trajan's Column depicting the bridge that Apollodorus built for Trajan.

Apollodorus was the foremost engineer of Imperial Rome in the 2nd century CE, when it was at the height of its power. This was a fortunate time to be an engineer. There was a long period of relative peace and prosperity known as the Pax Romana, and Rome's emperors commissioned ambitious building programmes as a statement of imperial power. Public buildings and monuments showcased the might and sophistication of Rome to the outside world. They also promoted the public image of the emperor to the people of Rome and the wider empire. This was why they needed skilled engineers like Apollodorus, who could make their grand designs reality.

Born in the city of Damascus, in what was then the Roman province of Syria, Apollodorus may have been educated there. If so, he would have been exposed to ideas and influences from the East and classical Arab culture, as well as from Rome and Greece. However, we can only speculate about Apollodorus' early life. He first appears in historical accounts as a military engineer, serving under the emperor Trajan. Engineering played a vital part in Roman military campaigns. Alongside a sword and spear, a shovel was part of the basic kit each soldier carried. It was regularly used for digging trenches and building temporary military camps while the legion was on the march. Legionaries were put at the disposal of military engineers such as Apollodorus to carry out more ambitious engineering works, such as building bridges, siege weapons, roads or permanent fortifications. Sometimes these engineering works played a decisive part in a military victory.

Apollodorus proved invaluable to Trajan during the emperor's second campaign against the Dacians, a people that ruled an area of eastern Europe north of the Danube river. As part of his invasion strategy, Trajan called on Apollodorus to construct a bridge across the Danube so that his legions could be rapidly deployed into enemy territory. Apollodorus rose to the challenge. First, he diverted the river so that wooden piles could be driven into the riverbed. These were used as the foundations for 20 hollow rectangular piers, which were then filled with stone and concrete, and dressed using long, flat Roman bricks and cement. Once these solid piers were in place, Apollodorus set wooden arches on top of them, which supported a deck of oak 15 m (49 ft) wide. On completion, the bridge was 1,135 m (3,723 ft) long. It was the longest arched bridge to have been built for more than a thousand years. Apollodorus managed a labour force drawn from several legions to complete the bridge in 105 CE. It was a tremendous achievement and played a significant part in Trajan's defeat of the Dacians the following year. It is thought that Trajan's final victory was sealed by the siege and overthrow of the Dacian capital, in which case Apollodorus would have played his part with siege weaponry and breaching the city's defences. Trajan returned to Rome in triumph, bringing with him the wealth of Dacia's gold mines – and his most trusted engineer Apollodorus.

Following an extended period of gladiatorial games to celebrate his conquest, Trajan embarked on a building programme in Rome to consolidate his power and commemorate his great victory in a more permanent way. Apollodorus was Trajan's chief engineer for these big-budget projects, funded by the spoils of the Dacian war. The first of Apollodorus'

BELOW: Reconstruction of a section of Trajan's Bridge over the Danube.

LEFT: A 19th-century painting showing Trajan's Column towering over the ruins of Trajan's Forum in Rome.

The Forum was completed around 112 CE, followed by the addition of Trajan's Column in 113 CE. This was another engineering triumph by Apollodorus, who built a column 30 m (98 ft) high, using 20 colossal cylinders of marble stacked atop a giant pedestal. A spiral staircase ran through the hollow centre of the column and its outer surface was covered with a spiralling frieze 190 m (623 ft) long, depicting Trajan's campaigns against the Dacians. Topped with a capital block bearing a statue of the emperor, the engineering logistics involved were challenging, requiring huge marble cylinders weighing more than 30 tonnes to be lifted into place with pulleys or cranes powered solely by human labour or animals.

Built to commemorate Trajan's military triumphs, Apollodorus' column became Trajan's funeral monument. An urn containing the emperor's ashes was placed in the base. Following Trajan's death, Apollodorus served Hadrian, but accounts suggest that he fell out with the new emperor and may even have been assassinated on his orders as a result.

commissions was the building of Trajan's Forum, the centre of political, commercial and religious life in Rome. It was a significant engineering challenge to prepare the site, which basically involved removing a large section of hillside. Labourers had to shift hundreds of thousands of cubic metres of rock and soil before building could even begin. Alongside the Forum, Apollodorus built Trajan's Market, a prototype for the shopping mall of the future and later constructed huge public baths.

As the chief engineer and architect of ancient Rome, Apollodorus reshaped the city with landmark new projects, as well as preserving its past with renovations to revered earlier Roman buildings. His prolific works represent a huge engineering achievement. Apollodorus' contributions to Roman engineering added to a legacy of structures that were rediscovered and drawn upon for inspiration and technical insight by later engineers, from the Renaissance to today.

ISMAIL
AL-JAZARI

'HE WAS A MASTER CRAFTSMAN, FULLY CONVERSANT WITH ALL BRANCHES OF HIS TRADE, CONSCIOUSLY PROUD OF HIS MEMBERSHIP OF THE TECHNICAL FRATERNITY. MORE RARELY, HE WAS A MASTER CRAFTSMAN WHO COULD WRITE, AND WHO HAS LEFT US AN ENGINEERING DOCUMENT OF THE FIRST IMPORTANCE.'

Donald Hill, chartered engineer and translator of Al-Jazari's *The Book of Knowledge of Ingenious Devices*

GREATEST ACHIEVEMENTS

THE BOOK OF KNOWLEDGE OF INGENIOUS DEVICES *c.* 1206

CASTLE CLOCK *c.* 1206

ELEPHANT CLOCK *c.* 1206

MUSICAL AUTOMATA *c.* 1206

ABOVE: Ismail al-Jazari.

Scattered across academic libraries and private collections around the world are fragmentary versions of a book entitled *The Book of Knowledge of Ingenious Devices*. The original was written by an Islamic scholar and mechanical engineer known as Al-Jazari at the start of the 13th century. His book contains details of how to construct 50 mechanical devices. These were machines Al-Jazari designed while in the service of a royal family of Upper Mesopotamia, an area that encompassed most of modern-day Iraq as well as parts of Syria and Turkey. Al-Jazari's devices range from practical tools, such as machines to raise water and various clocks, through to elaborate mechanical wine dispensers and musical automata.

All we know about the life of Al-Jazari is contained in a handful of details from the introduction to his book. He gets his name from the region where he was born – Al-Jazari, the northernmost of three provinces of Upper Mesopotamia. This fertile and economically prosperous area lay between the rivers Tigris and Euphrates. It was ruled during this period by the Abbasid Caliphs and was part of a large Islamic empire. The Abbasids fostered learning and scholarship, establishing libraries and promoting the translation of texts from Greek, Persian, Indian and Chinese. This made ideas from a wide range of traditions and disciplines available to Islamic scholars.

Al-Jazari was in the service of the king of Diyar Bakr in Upper Mesopotamia when he was instructed to write his book. By this time, he had been court engineer for 25 years. Al-Jazari was working during an extended period of peace at the height of the Islamic Golden Age. Libraries flourished and scholars were drawn to these centres of learning. Scholarship was valued, as was the preservation of knowledge in books. The king asked Al-Jazari to compile a selection of his mechanical devices in a book so that his chief engineer's knowledge would not die with him. In fact, Al-Jazari completed his task in 1206, just a few months before he died.

Before writing his book, Al-Jazari embarked upon a study of earlier texts to incorporate what could be learned from them into his work. Likely sources include the Greek engineer Heron of Alexandria, a renowned builder of automata, as well as an expert on

BELOW: One of Al-Jazari's designs for a water clock. The figure on the top is an automaton that points to the hours.

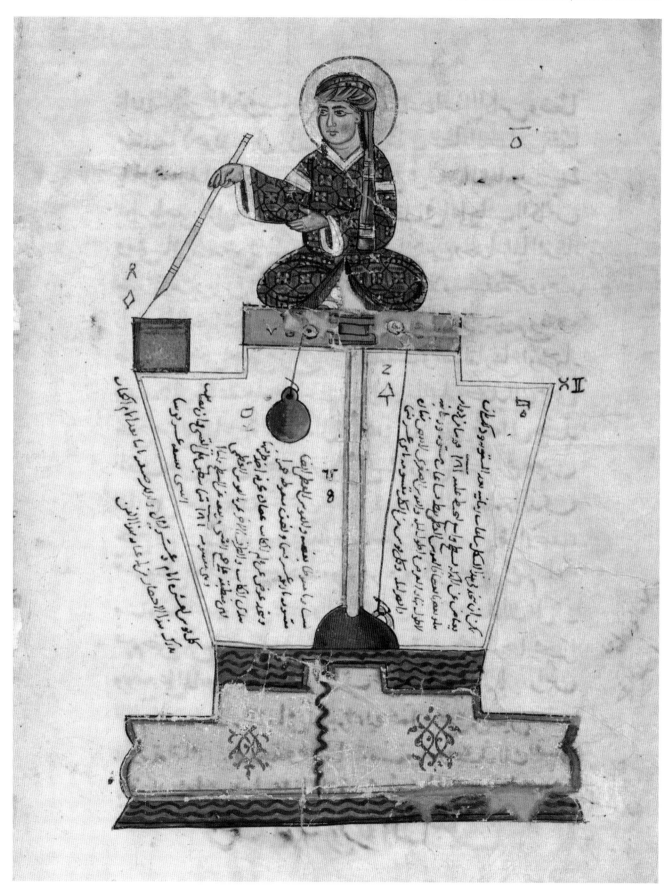

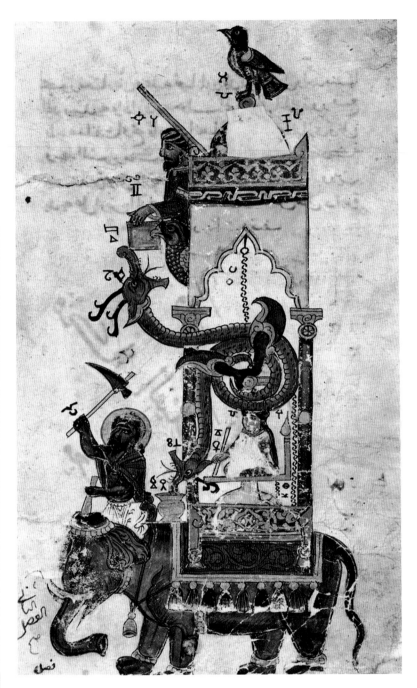

OPPOSITE: The Castle Clock from Al-Jazari's *The Book of Knowledge of Ingenious Devices.*

texts, making technical improvements and ensuring his des-criptions were intelligible to future engineers. Al-Jazari's meticulous ap-proach made his book popular in its day. It has been of lasting value to historians of engineering ever since.

Some of the devices Al-Jazari describes in his book, look at first sight to be frivolous novelties designed to amuse a king and his guests. But amid the detailed descriptions of musical automata and wine-dispensing novelties a clear engineering methodology unfolds. Al-Jazari's book is a practical manual, with step-by-step instructions for building each machine.

Among the designs are a number of innovations that were ahead of their time. Al-Jazari's Castle Clock uses an early camshaft to turn rotational motion into up-and-down motion. This and the early crankshaft described in one of his water pumps would go on to be an important feature of car engines. The Castle Clock is also considered by some to be an early mechanical computer. Likewise, some of the musical automata, with their mechanical drummers or singing birds, may have been programmable, anticipating later developments in robotics and computing.

Al-Jazari wrote *The Book of Knowledge of Ingenious Devices* with a uniquely engineer-focused clarity, encompassing both materials and manufacturing techniques. His painstaking commitment to providing the fullest detail extended to painting miniature illustrations of his devices. It is these colourful images, as much as his text, that provide us with such a vivid window on the engineering achievements of the medieval Arab world.

water clocks whose work was ascribed to Archimedes at that time. Al-Jazari would also have been familiar with the Bānū Mūsā brothers, three Persian scholars who had produced a compendium of a hundred ingenious devices four centuries earlier.

Al-Jazari's researches were not uncritical of these earlier texts. He often found their instructions incomplete or that the devices described were impractical to build, lacking in technical rigour. Al-Jazari was a hands-on engineer and diligently set about addressing the shortcomings of previous

وهذه صورة ما وصفته قابضة

FILIPPO
BRUNELLESCHI

GREATEST ACHIEVEMENTS

IL BADALONE
A boat for transporting marble, 1427

THE DOME OF THE CATHEDRAL OF SANTA MARIA DEL FIORE
Florence, 1436

THE FOUNDLING HOSPITAL
Florence, 1419–45

THE BASILICA OF SAN LORENZO
Florence, completed in 1442

THE BASILICA OF SANTO SPIRITO
Florence, 1434–46. The façade for the basilica was not completed until 1482, many years after Brunelleschi's death.

ABOVE: Filippo Brunelleschi.

'BOTH THE MAGNIFICENT DOME OF THIS FAMOUS CHURCH AND MANY OTHER DEVICES INVENTED BY FILIPPO THE ARCHITECT, BEAR WITNESS TO HIS SUPERB SKILL.'

Epitaph for Filippo Brunelleschi from interior of the Basilica di Santa Maria del Fiore, Florence.

During the Middle Ages, the writings of ancient Greek and Roman authors that had been preserved and studied during the Islamic Golden Age began to filter into Europe. Starting in Italy, this revival of classical knowledge fuelled the Renaissance: a period of social change, artistic creativity and technical advancement that occurred during the 15th and 16th centuries. In antiquity there had been no division between the arts and sciences. During the Renaissance, the interplay between the arts, science, architecture and engineering reached new heights with the achievements of multi-talented polymaths, including the Florentine architect and engineer Filippo Brunelleschi. His crowning glory was the dome of Santa Maria del Fiore cathedral in Florence, a structure that embodies a union of engineering, architecture and artistry.

We know little of Brunelleschi's early life. He was born in 1377, in the Italian city of Florence. He came from a wealthy family and had a good education, but instead of following his father into the legal profession, young Brunelleschi became an apprentice of the silk merchant's guild in Florence. One of seven major guilds in the city, it provided a diverse training that embraced art and artisanal disciplines, including

ABOVE: Brunelleschi's colossal dome crowns the Basilica of Santa Maria Del Fiore in Florence.

metal-working, bronze sculpture and making jewellery.

At the age of 22 Brunelleschi became a fully fledged master goldsmith. However, architecture and sculpture in bronze also occupied Brunelleschi early in his career, as did the making of clocks, if we are to believe the account of an early biographer. In 1401, Brunelleschi entered a competition to design the bronze door panels of the Baptistry of San Giovanni in Florence. He lost out, coming second to Lorenzo Ghiberti, a rival goldsmith and artist. Unhappy at being pipped to the post, it is thought that Brunelleschi travelled to Rome with the artist Donatello and sought distraction from his wounded pride with an intense study of classical

OPPOSITE: A cross-section of Brunelleschi's design for the dome.

RIGHT: Brunelleschi's revolving crane.

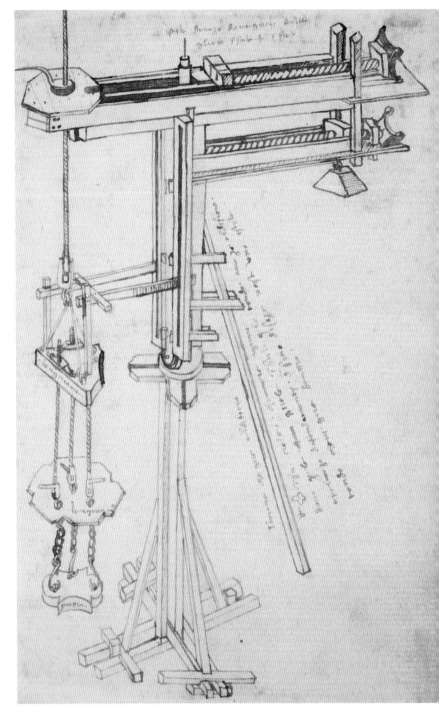

architecture. Nevertheless, even though he was the runner-up, the competition brought Brunelleschi publicity and the attention of wealthy patrons.

In 1415, Brunelleschi made a significant contribution to the visual arts, rediscovering linear perspective: a method of drawing that enabled an artist to convincingly render three-dimensional objects on the flat surface of a picture. It influenced many Renaissance artists, but another likely use of the technique must surely have been that it allowed Brunelleschi to communicate his engineering and architectural plans both to clients and his workforce. Sadly none of Brunelleschi's own drawings have survived, probably due to his need to protect his designs and techniques from the prying eyes of rival artist-engineers and their workshops.

In 1418, Brunelleschi entered another public competition. This time it was for a much bigger and more prestigious commission: building the dome to complete the cathedral of Santa Maria del Fiore, which was nearing completion after more than a century of building. The challenges of constructing a dome over its huge octagonal east tower had frustrated earlier architects. Once again, Brunelleschi was up against Ghiberti, but in 1420 he finally won the top job with his plan to build a double-shelled dome from bricks. Brunelleschi's plan was ingenious and economical, avoiding the need for a massive wooden framework to support the dome during construction. Instead, he devised a method to lay about four million bricks in a zigzag pattern between stone beams so that they would hold themselves up. He also designed and built a series of highly manoeuvrable oxen-driven hoists and rotating cranes that were at the heart of the construction site. Experts estimate that these lifted around 13,000 tonnes of bricks and marble up into the dome every day.

Brunelleschi's success in building the dome combined innovative engineering solutions with meticulous planning and project management. It took him just 15 years to raise the dome from its base, 54 m (177 ft) off the ground, to the circular opening at its top, about 33 m (108 ft) above. Brunelleschi's mastery of the complex logistics as well as the technical challenges of constructing the dome are why many consider him the world's first modern engineer. While the dome was being built, Brunelleschi also worked on numerous

other architectural and engineering projects. His first architectural commission, awarded in 1419, was to design the Foundling Hospital in Florence. The churches of San Lorenzo and Santo Spirito followed, cementing his growing reputation for engineering excellence and artistry.

Beyond architectural projects, Brunelleschi's skills as a mechanical engineer found other outlets. In 1425, he was granted what is thought to be the world's first engineering patent for a boat designed to lift and transport heavy marble slabs to Florence from the quarries at Carrara. His design of lifting and hoisting mechanisms may have drawn on a knowledge of gearing learned from clock-making. Sadly Brunelleschi's vessel, named *Badalone*, foundered on the Arno river during its maiden voyage and is thought to have sunk along with its 100-tonne load of marble. Trailblazing engineering meant that Brunelleschi's career had its fair share of failures. Another example occurred in 1428 when Brunelleschi was engaged as a military engineer to aid Florence in its war with nearby Lucca. The idea was to isolate the city of Lucca by diverting the Serchio river around it, but the plan backfired and flooded the Florentine camp instead.

The dome of Santa Maria del Fiore was completed in 1436. Towering almost 90 m (295 ft) over the cathedral floor, it had taken about 37,000 tonnes of brick and stone to complete and was a testament to Brunelleschi's double-dome design and the accuracy of his load-bearing calculations. Brunelleschi had to compete against Ghiberti one more time before his design for a lantern to top the dome was accepted. Sadly Brunelleschi died in 1446, just after its construction had started.

In 1471, the artist-engineer Andrea del Verrocchio installed a gilded copper ball on top of Brunelleschi's lantern. With him he brought a young apprentice who was fascinated by Brunelleschi's lifting machines and sketched their workings. The apprentice was Leonardo da Vinci. He was just one of the artist-engineers that would be inspired by the pioneering work of Filippo Brunelleschi.

LEFT: A model of Brunelleschi's giant boat *Il Badalone*. Brunelleschi patented his cargo-carrying boat design – this was the first engineering patent to ever be granted.

MARIANO DI JACOPO:
IL TACCOLA

'INGENUITY IS WORTH MORE THAN THE STRENGTH OF BUFFALOES.'

Mariano di Jacopo, Taccola: *De Machinis*

Until relatively recently, the incredible explosion of arts and literature during the Renaissance tended to overshadow the technical achievements of the period. Now it is more widely recognized that many of the talents who created masterpieces of painting and sculpture were also responsible for impressive achievements in architecture and engineering. Some commentators have referred to this parallel strand of technological development as the 'Renaissance of machines' and have described the rise of the multi-skilled artist-engineer as its driving force.

Mariano di Jacopo, widely known by his nickname 'Taccola' (crow), was one of these hybrid Renaissance talents, being both an artist and an innovative engineer. Like his slightly older contemporary Brunelleschi, Taccola helped to progress the transition from the medieval world into the Renaissance. Though Taccola worked on real-world engineering projects, his two highly-illustrated treatises: *De Ingeneis* (*About Engines*) and *De Machinis* (*About Machines*) are perhaps his real legacy. A mix of practical engineering plans and more fanciful blue-sky thinking stretches across these four volumes. All his ideas are made immediate and intelligible thanks to Taccola's accomplished pen and ink drawings. These texts represent a true fusion of the artist and engineer. They brought renown to Taccola, his home city of Siena and his patrons.

Taccola was born in the Italian city of Siena in 1382. We know little of his formative years and education, but as a young man Taccola trained in the workshop of the sculptor Jacopo della Quercia. This is where he must have learned to draw as well as how to carve and work with stone. In 1408, Taccola worked as a sculptor on the choir of Siena's cathedral. His career then diversified for a spell, as Taccola took up a job within the administration of Siena's flagship academic institution the Casa della Sapienza. He subsequently worked as a notary, but Taccola continued to soak up

BELOW: Manuscript page with Taccola's intricate technical drawings of a variety of boats and machines.

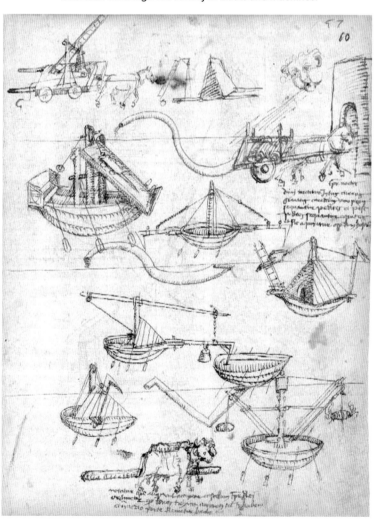

BELOW: Muscle power and two giant bellows powered Taccola's design for a hydraulic water pump.

the knowledge and new ideas flowing into Siena.

Taccola became increasingly fascinated by machinery, probably as a result of seeing versions of classical scientific texts or later Islamic works on machines brought to Siena by visiting scholars. He put his growing knowledge of mechanics into practice as an engineer on construction projects in Siena. The city, which was at the heart of the Sienese Republic, was rapidly expanding during this period. Siena had the wealth to finance a range of civic works and it needed the expertise of engineers such as Taccola, and a little later Francesco di Giorgio. Not only could they help to maintain and extend Siena's infrastructure, but they could help to defend it from

ABOVE: Below the streets of Siena, a network of ancient aqueducts known as the 'Bottini' still supplies the city with water today.

warring competitors by constructing fortifications and making new war machines and weapons.

Siena was well-provided with minerals and resources for its developing industries, but it had a major problem in maintaining an adequate supply of water. Earlier engineers had built an elaborate 25-km (15.5-mile) network of underground aqueducts known as 'bottini' to address the problem. However, this system needed maintenance by skilled engineers to keep the water and the city's showcase fountains flowing. Given the importance of water in Siena, it is not surprising that Taccola made hydraulics his engineering specialism. His interest in this area is clear from his writings, which contain all kinds of water management devices, including siphons, screw pumps and sluice gates that could control the speed of water flow for mills. When the future emperor of Hungary, Sigismund visited in 1432, Taccola pitched himself as both a hydraulic engineer and a painter.

Sigismund became a patron as a result. It seems Taccola was never shy about his talents. He opens his book *De machinis* by describing himself as the 'Archimedes of Siena'!

Everyday engineering work, such as supervising road building, occupied Taccola for part of his career, but it was on paper that he was free to give his unique blend of artistic creativity and technical knowledge free rein. In the latter part of his life he compiled the pages of ideas he had accumulated over a long career into two books: *De Ingeneis* and *De Machinis*. They contain a mix of the tried-and-tested alongside more speculative ideas. There are mathematically sound methods for surveying sites or boring tunnels from either side of a hill to meet in the middle, and designs for a thoroughly practical rock-hurling trebuchet weapon. But there are also thinking-out-loud ideas such as an over-elaborate pulley-based device for fishing or an almost-certainly impractical method for floating mounted cavalry soldiers across a river using inflatable leather bags.

The closing pages of Taccola's *De ingeneis* bear notes and drawings by the younger Sienese engineer Francesco di Giorgio. Following Taccola's death in 1453, di Giorgio would continue to combine text and illustrations in a manner similar to Taccola to communicate his engineering ideas. Francesco's treatise on architecture would later influence the ultimate artist-engineer: Leonardo da Vinci.

BELOW: A series of levers and pulleys can be clearly seen at work in Taccola's design for a crane.

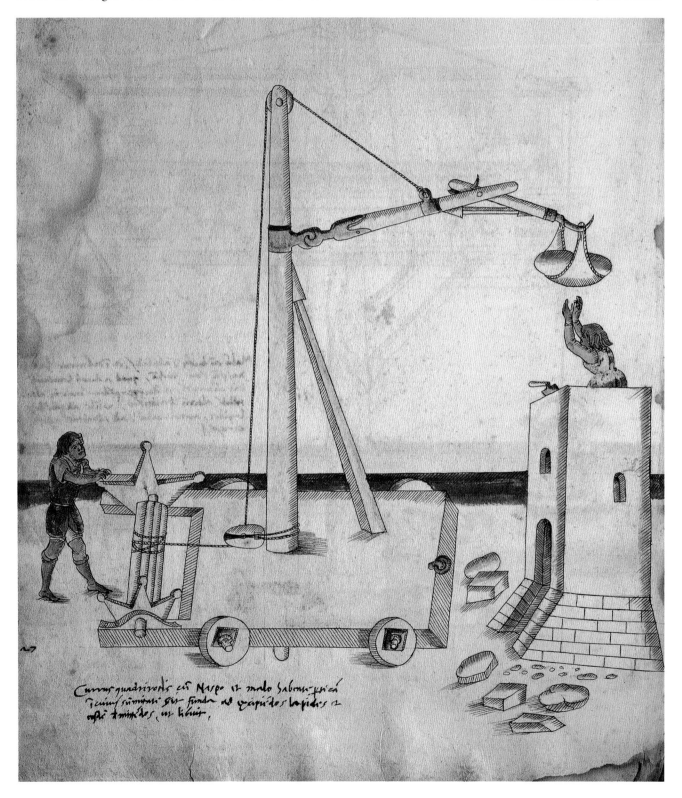

LEONARDO DA VINCI

GREATEST ACHIEVEMENTS

GRAN CAVALLO
Leonardo da Vinci began work on the Sforza monument in 1482 but was forced to abandon the project after the war with France began in 1494.

DESIGNS FOR FLYING MACHINES c. 1488–9

THE VITRUVIAN MAN c. 1490
This classic study of anatomy represented just one of da Vinci's many interests.

THE LAST SUPPER c. 1490s

MONA LISA c. 1516

ABOVE: A portrait of Leonardo da Vinci.

'THOUGH HUMAN INGENUITY MAY MAKE VARIOUS INVENTIONS, IT WILL NEVER DEVISE AN INVENTION MORE BEAUTIFUL, MORE SIMPLE, MORE DIRECT THAN DOES NATURE...'

Leonardo da Vinci, *Codex Windsor*

Leonardo da Vinci is probably the most famous artist of the Renaissance: a household name, remembered for iconic paintings such as the *Mona Lisa* (*La Gioconda*) and *The Last Supper*. But art was just one of Leonardo's many talents, which ranged across science, anatomy, mathematics, architecture and engineering. Leonardo himself would have seen no distinction between these fields of knowledge and skill. Rather, all were encompassed within a humanist-influenced study of nature through observation and experiment.

Science and mathematics served Leonardo's art, helping him to deal with technical challenges such as how to cast a giant bronze equestrian statue or with the tricky perspective required to create a painting of *The Last Supper,* high on the refectory wall of Santa Maria delle Grazie in Milan, Italy. In a reciprocal way, his drawing skills meant Leonardo could visually record scientific observations and ideas, as well as produce detailed engineering and architectural designs.

Like Taccola and di Giorgio, Leonardo was an artist-engineer with a technical knowledge and creative skill that allowed him to speculate, design and invent on paper. Around 7,000 pages of Leonardo's illustrated notes, all written in reversed 'mirror writing' have survived. These were originally loose sheets

ABOVE: Leonardo da Vinci's painting of *The Last Supper* in the refectory of Santa Maria delle Grazie, Milan.

that he covered with notes and drawings wherever he was working. He may have used some pages to pitch ideas to wealthy patrons. Others record observations or scientific studies. Later, these manuscript pages were compiled and bound into notebooks. Today they are a treasure trove of ideas and inventions, and clearly reveal how central engineering was to Leonardo's career.

Leonardo was born in 1452 in the town of Vinci, amid the Tuscan hills of central Italy. He was the illegitimate son of Piero da Vinci, a notary who worked in Florence and Caterina Lippi, a young woman from a poor farming family. As an infant he was raised by his mother on a farm in the countryside. Then, around the age of five, Leonardo was brought to live with his uncle and his grandparents in the family house in Vinci. However, as Leonardo was not a legitimate heir, he didn't receive the standard education with its emphasis on bookish rote-learning based upon classical texts. Ironically, this may have helped to

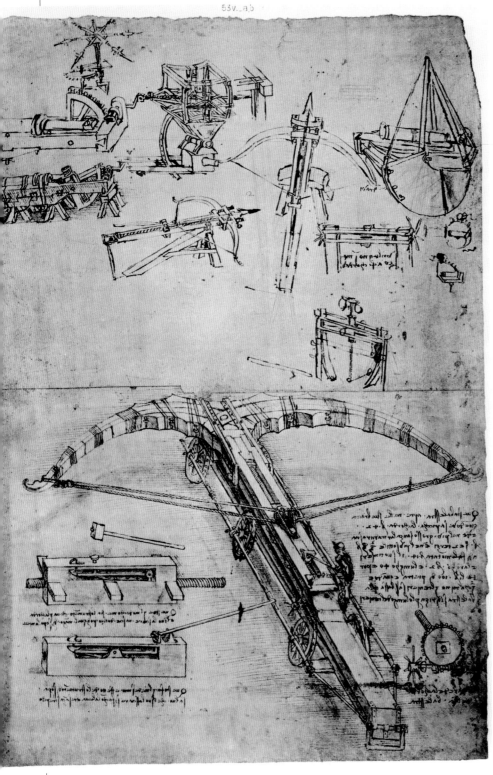

the age of 14, his father found Leonardo an apprenticeship in the studio of the renowned Florentine artist Andrea del Verrocchio. Primarily a sculptor, Verrocchio was also a painter and a goldsmith. Verrocchio's busy workshop handled a variety of commissions and Leonardo would have picked up a wide range of skills and knowledge, from drawing, sculpture and painting through to metalwork and casting, mechanical engineering, chemistry and carpentry.

The apprentice Leonardo honed his artistic talents under Verrocchio's supervision. According to one famous account, Leonardo was asked to add an angel to one of Verrocchio's paintings. The result was so stunningly accomplished that Leonardo's master apparently gave up painting afterwards. Leonardo also got first-hand practical experience of engineering. He assisted Verrocchio in installing a golden ball on top of Brunelleschi's dome on Santa Maria del Fiore in Florence. While on the cathedral worksite, Leonardo avidly studied the machinery used in the dome's construction. Leonardo's drawings of Brunelleschi's revolving crane and an elevator-type hoist used to lift materials up into the dome are preserved in his notebooks. Leonardo continued to work with Verrocchio after he had qualified and started to get his own commissions. He also started to gain a reputation for not finishing his projects, possibly as a result of Leonardo's many other interests distracting him.

nurture the young Leonardo's freedom of thought and self-reliance. He would later make a virtue of being what he described as an 'unlettered man', anticipating Galileo with a reliance on observation and reasoning rather than received wisdom from books.

Following the death of his stepmother, Leonardo went to live with his father in Florence. Here, at

ABOVE: Ludovico Sforza.

BELOW: Leonardo's designs for mechanized war machines included a chariot with rotating blades and a prototype for an armoured vehicle that anticipated the modern tank.

In 1482, at the age of 30, Leonardo was sent from Florence to Milan by Lorenzo the Magnificent. Leonardo was to present a silver lyre that he had made to the Duke of Milan as part of a diplomatic mission to smooth over conflict between the two city states. Leonardo seized his opportunity and pitched for work as a military engineer with Ludovico Sforza, who was soon to replace his father as Duke of Milan. Leonardo wrote a letter to Sforza that detailed the engineering skills that he could bring to a military campaign, such as building bridges and tunnelling. He goes on to list a wide variety of weapons that he could make, including armoured vehicles, artillery and other siege weapons. Leonardo closes his letter citing abilities in civil architecture and hydraulic engineering that would be of use in times of peace. He only mentions his artistic skills as a brief afterthought!

Ludovico was impressed and Leonardo got the job. During this period the status of artist-engineers in Italy was at an all-time high. Rival patrons from wealthy families and rulers from further afield competed to bring their talents into their court. Most

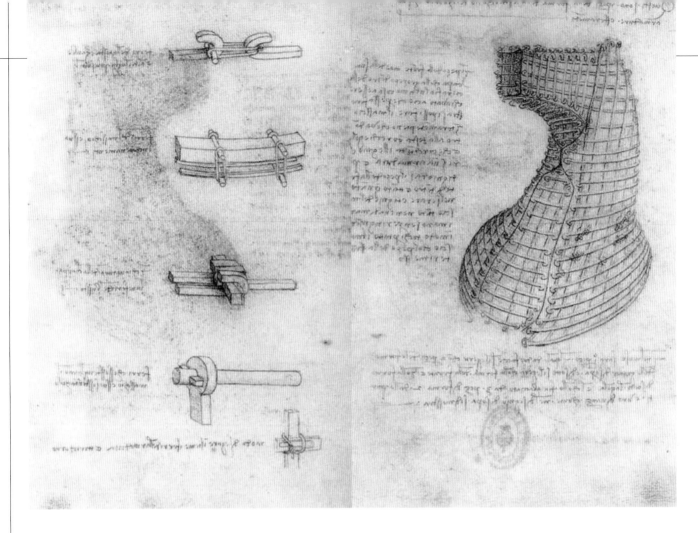

of Leonardo's military inventions would never get built, but he provided the special effects for court entertainments. He was also commissioned to create a giant bronze statue of Ludovico Sforza's father on horseback. It would take Leonardo several years to address the technical challenges of casting such a large statue. Leonardo's procrastination earned him the mocking scorn of rival artist Michelangelo. He did get as far as making a colossal clay model, dubbed the 'Gran Cavallo', and around 70 tonnes of bronze were bought in, ready for Leonardo to start casting. Sadly, events overtook the project. Conflict with France saw the bronze sent off to make cannons to repel Charles VIII's forces. Then Milan was captured by the French, forcing Leonardo to flee to Venice. He abandoned his huge clay model of the 'Gran Cavallo'. It was later used for target practice by occupying French troops.

Following his exit from Milan, Leonardo moved around, mixing studies with service as a military and civil engineer for powerful rulers in Venice, Mantua, Florence and Rome. He even briefly returned to Milan in 1506 at the invitation of the French governor. Increasingly Leonardo focused on his own studies and interests, covering page after page of his notebooks with everything from detailed drawings of human anatomy, through to studies of various types of screw mechanisms or incredible ahead-of-their-time inventions.

In 1513, at the age of 61, Leonardo was invited by Pope Leo X to live at the Vatican in Rome. It was here that he met the king of France, Francois I, a great patron of the arts who admired Leonardo's work. He offered the now elderly artist-engineer a role in his court. Leonardo moved to the royal manor house at Clos Lucé in France. He was granted a pension and had the freedom to work as he pleased. Leonardo died there in 1519, reportedly in the king's arms. He left behind the unfinished portrait of the *Mona Lisa* that he had carried with him wherever he went, plus thousands of notebook pages. Though many of Leonardo's plans, from the ideal city through to flying machines would never get further than these drawings, his boundless creativity still inspires engineers today.

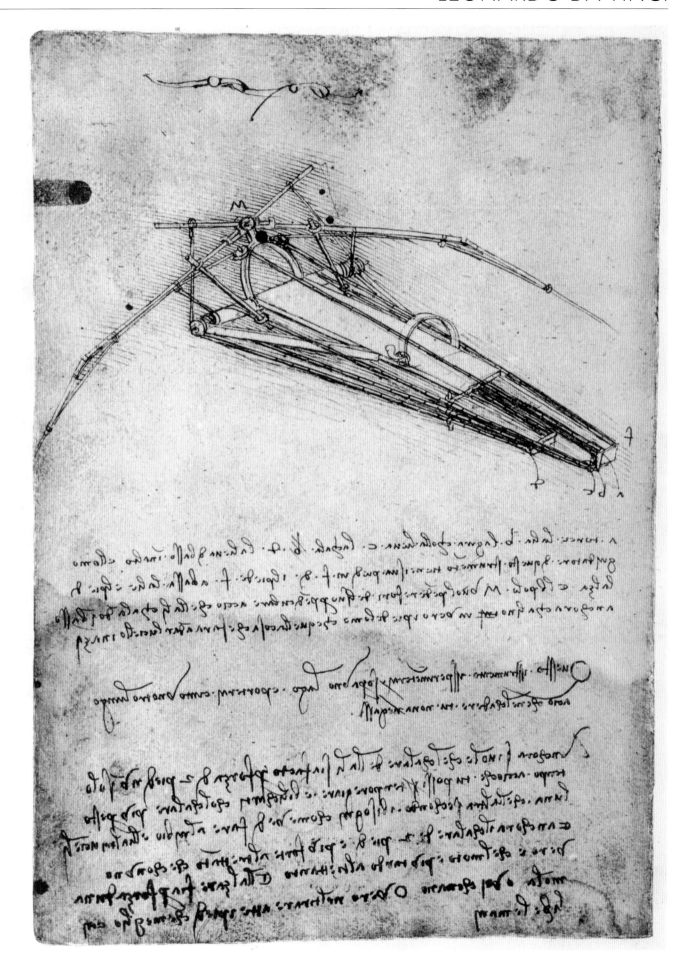

CORNELIS
DREBBEL

'NOT ONLY FROM HIS HANDS, BUT ALSO FROM HIS MIRACULOUS BRAIN COMES, WHAT I CALL THE STANDING TELESCOPE. HAD DREBBEL MADE NOTHING MORE THAN THIS MIRACULOUS TUBE DURING HIS LIFE, HE WOULD HAVE ACQUIRED IMMORTAL FAME.'

Constantijn Huygens on Drebbel's microscope

GREATEST ACHIEVEMENTS

THE 'PERPETUUM MOBILE' CLOCK 1598

COMPOUND MICROSCOPE EARLY 17TH CENTURY

OAR-POWERED SUBMARINE 1620

DREBBEL'S CIRCULATING OVEN 1620s
This self-regulating oven included an early version of a thermostat.

ABOVE: Cornelis Drebbel.

The Dutch inventor and engineer Cornelis Drebbel is chiefly remembered for building the world's first working submarine. However, this impressive technical feat has overshadowed his other contributions to engineering across areas such as hydraulics, chemical engineering, feedback control mechanisms and optics. His career spanned the transition from the Renaissance to what has been called the 'Age of Reason', when the focus of scientific and technological innovation shifted from Italy to northern Europe. It was a time of exploration, empires and expansion and the skills of engineers such as Drebbel were in demand.

Drebbel was born in Alkmaar in the Netherlands in 1572. Following a classical-based education, Drebbel became an apprentice engraver and map-maker. He studied under the famous engraver, painter and humanist philosopher Hendrik Goltzius in Haarlem. Goltzius was also interested in alchemy, which combined a mysterious and secretive philosophy with laboratory work that anticipated the science of chemistry. Among other things, alchemists were interested in turning base

ABOVE: Drebbel's perpetual motion clock cleverly used changes in atmospheric pressure to keep its spring wound up.

metals into gold and their experiments laid the foundations for both chemistry and its practical application in chemical engineering. Goltzius may well have passed on some of his knowledge of alchemy to the younger Drebbel, who later proved an adept chemist, making fireworks, dyes and explosives.

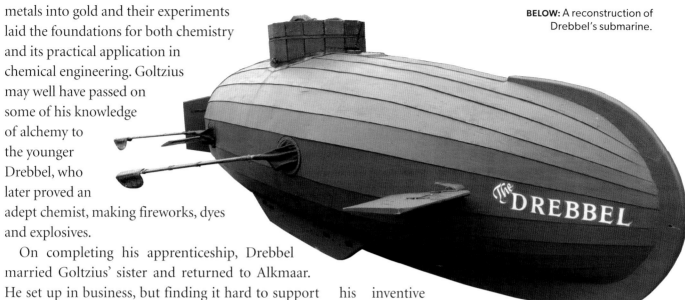

BELOW: A reconstruction of Drebbel's submarine.

On completing his apprenticeship, Drebbel married Goltzius' sister and returned to Alkmaar. He set up in business, but finding it hard to support his household through engraving, he branched out into engineering. In 1598, Drebbel was awarded two patents. One was for a pump used to raise fresh water to a public fountain in the town of Middelburg. The other was for a clock whose spring-powered mechanism never needed winding up. Described as a perpetual motion machine, it caught the public imagination and gave Drebbel's engineering credentials a boost. Drebbel's clock cleverly used changes in atmospheric pressure to keep its spring from running down.

The boom in scientific research in northern Europe provided further employment for Drebbel. He turned his dexterity as an engraver to the more lucrative business of making scientific instruments. He learned how to grind lenses for telescopes and microscopes and also wrote several scientific books. With his reputation growing, he moved to England to become King James I's engineer. In practice, this meant a good deal of supplying ingenious special effects for court entertainments or fireworks and it wasn't long before Drebbel's restless talent looked for other outlets. Eventually, he offered his services to Holy Roman Emperor Rudolf II, whose court was in Prague and who shared an interest in alchemy. Drebbel moved to Prague in 1610. Unfortunately, Rudolf II was deposed by his brother, Archduke Matthias, the following year. As a result, Drebbel spent a year in prison before being allowed to return penniless to England.

Once again, the ever-industrious Drebbel put his inventive brain to work. He built a lens-grinding machine and developed a much higher-magnification compound microscope using two lenses. Using his skills in chemistry, Drebbel developed a brighter and more colourfast red dye. He also invented a type of thermostat to control the temperature of a more fuel-efficient stove. This was an important engineering innovation in its use of an automated feedback control mechanism. Drebbel then went on to design an incubator to help hatch chicken eggs by keeping them at a constant temperature.

Having worked his way back into the court of King James I, Drebbel built and demonstrated his oar-powered submarine in the murky waters of the River Thames in 1620. It wasn't his invention, but Drebbel had the engineering skills to realize a design published by William Bourne in 1578. Two further prototypes were built, but although Drebbel's submarine worked perfectly well, it proved too far ahead of its time for the British Royal Navy. However, they were sufficiently impressed to take on Drebbel as a naval engineer. Sadly Drebbel's career took a downturn after this. Following the failure of his technical innovations to make an impact during a naval campaign against the French, Drebbel lost his job. He ended up running an ale house in London and died in relative obscurity in 1633. It was a disappointing end to the life of an engineer whose genius and prodigious output across many fields has seen him later compared to Thomas Edison.

CHRISTIAAN
HUYGENS

'THE WORLD IS MY COUNTRY; SCIENCE IS MY RELIGION.'

Christiaan Huygens

GREATEST ACHIEVEMENTS

DISCOVERED SATURN'S LARGEST MOON, TITAN
1655

PENDULUM CLOCK 1657

OBSERVATION OF SATURN'S RINGS 1659

POCKET WATCH 1675

ABOVE: Christiaan Huygens.

The Dutch polymath Christiaan Huygens was one of the leading scientists of the 17th century, as well as being a mathematician, astronomer, inventor and engineer. Huygens made significant contributions to science with a wave theory of light, as well as important studies regarding forces and the mathematics of probability. As an astronomer, he built an improved telescope and was the first to discover that the planet Saturn was encircled by rings. Huygens was also a skilled mechanical engineer. He developed the world's first pendulum clock and made another landmark contribution with a design for a pocket watch with a balance spring.

Huygens was born in 1629 into a wealthy family in the bustling city of The Hague, which was a major centre for government and legal administration. Christiaan's father, Constantijn Huygens, travelled frequently on diplomatic work and came into contact with intellectuals such as the Italian scientist Galileo and the French thinker René Descartes, who Christiaan met as a boy. As a consequence, Christiaan grew up surrounded by the ideas that were fuelling the Scientific Revolution. The young boy's intelligence and talent for mathematics were such that Christiaan's father called his son 'my Archimedes'.

The young Huygens was educated at home by private tutors, before enrolling at the University of

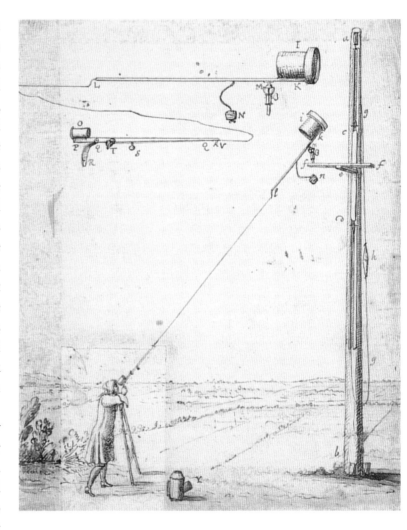

ABOVE: Huygens' aerial telescope.

Leiden to study law and mathematics in 1645. Two years later he switched to a new college in Breda, finishing his studies in 1649. On graduating it was expected that he would follow his father into a career in the diplomatic service, but Christian did not take to this kind of work and preferred to pursue his passion for mathematics. Being from a wealthy family, Christiaan was able to return home to the Hague where he continued his studies independently and exchanged ideas with other intellectuals through extensive correspondence.

Alongside his mathematical researches, Huygens became interested in optics and astronomy. The Netherlands was at the forefront of lens-making for spectacles and optical scientific instruments. Together with his older brother, Constantijn, Huygens set out to learn the painstaking skill of grinding and polishing lenses. This led him to build a machine to help produce large and accurately shaped lenses. Huygens then used these large lenses to construct a more powerful type of telescope. When he used his new telescope to observe Saturn, Huygens first discovered a moon orbiting the planet in 1655. It was later named Titan. Then in 1659 he was able to see and describe the rings of Saturn for the first time. Galileo had first spotted Saturn's rings in 1610, but had been unable to make out what they actually were as his telescope lacked the magnification. Huygens' discoveries had been made possible by combining his talents in engineering and optical science. His discoveries brought fame and the French king Louis XIV asked Huygens to join the prestigious Académie des Sciences. Huygens moved to Paris in 1666 where he continued his work in astronomy and optics.

Astronomical observations required precise timekeeping. The expansion of global trade had also made new demands on the accuracy of timepieces as they were crucial for the navigation of ships. Huygens picked up on Galileo's 1602 observations about the regular motion of pendulums and how the time taken for one swing was the same, no matter how long the swing was. Galileo realized that this property could be used to make a clock, but the Italian scientist never managed to turn his idea into a working prototype. It was left to Christiaan Huygens to practically realize the idea with his design for a pendulum clock in 1657. Built by local clockmaker Salomon Coster, Huygens' pendulum clock was a massive step forward in timekeeping accuracy from the spring-driven mechanical clocks of the day. They were the most accurate timepieces until the quartz clock was invented hundreds of years later.

ABOVE: A balance wheel for a clock.

Huygens would make one other innovation in clock engineering: the balance wheel. This was a weighted wheel with a spiral-shaped spring to rotate it first one way and then the other. Basically it did the same job as a pendulum, helping to regulate the movement of the mechanism. English scientist Robert Hooke had independently come up with the idea some years earlier, but it was Huygens with his skill in precision engineering who first presented a workable design. He used the idea for a pocket watch that he patented in 1675.

The pressing challenge of the time was to create an accurate timepiece for ships to help them to navigate. Huygens devoted a good deal of time in the 1660s to building prototypes and sending them off on sea trials. However, the rolling motion of the sea affected the motion of the pendulum and the clock failed to keep time with sufficient accuracy.

Ill health affected Huygens throughout his life and particularly in his latter years. He returned to the Netherlands from Paris in 1681 and continued to work and publish. He also travelled to lecture and met other scientists such as Isaac Newton, whose work would later overshadow Huygens' achievements. Huygens died in 1695 after a lifetime of ground-breaking work in science and engineering.

ROBERT HOOKE

'THE BUSINESS AND DESIGN OF THE ROYAL SOCIETY IS TO IMPROVE THE KNOWLEDGE OF NATURAL THINGS AND ALL USEFUL ARTS, MANUFACTURES, MECHANIC PRACTICES, ENGINES AND INVENTIONS BY EXPERIMENTS.'

Robert Hooke

GREATEST ACHIEVEMENTS

VACUUM PUMP 1655

HOOKE'S LAW 1660

ELECTED AS A FELLOW OF THE ROYAL SOCIETY 1663

MICROGRAPHIA 1665

APPOINTED SURVEYOR TO THE CITY OF LONDON 1666
After the Great Fire of London, Hooke helped to rebuild the city along with Sir Christopher Wren.

ABOVE: Robert Hooke.

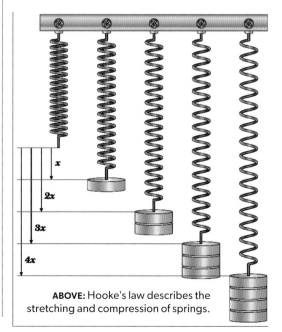

ABOVE: Hooke's law describes the stretching and compression of springs.

The Scientific Revolution sparked off by thinkers such as Francis Bacon and René Descartes gathered pace through the 17th century. The study of the natural world and the mechanisms governing it by 'natural philosophers' laid the foundations of modern science with an emphasis on experiment and observation. This period also saw the rise of scientific societies and academies throughout Europe, including the Royal Society in London. It was here that Robert Hooke emerged as a major natural philosopher. Hooke made discoveries with microscopes and telescopes, produced scientific theories and speculated on everything from physics to fossils. He then went on to use his technical and organizational skills to rebuild the city of London with Christopher Wren after the Great Fire of London in 1666.

Robert Hooke was a skilled engineer as well as a scientist, though engineering had yet to become a separate discipline. In fact, it was Hooke's mechanical genius that gave him the chance to work in a field dominated by wealthy gentlemen scientists, despite his modest background. The natural philosophers needed custom-built

apparatus and instruments for their experiments. Their workshops and proto-laboratories employed teams of artisans to produce the kit they used. This led to a development of engineering practices in parallel with the drive for scientific discovery. In an era of flourishing trade and growth, engineers were also employed in turning new scientific discoveries into commercially useful technologies and devices. Hooke was a polymath who bridged the worlds of science and engineering and he flourished as a result.

Robert Hooke was born in 1635 at Freshwater, a coastal village on the Isle of Wight. His father was curate of the local church. Robert was the youngest child in the family. He was often ill, so was largely home-schooled by his father. The young Hooke showed an early talent for drawing and for making models. As a boy, he is said to have built a working replica of a clock's mechanism out of wood, as well as a model battleship with cannons that actually fired. Hooke was just 13 when his father died, leaving him a small legacy. With this money he went to London to start an apprenticeship with the renowned portrait painter Sir Peter Lely. However, Hooke's artistic ambitions proved short-lived. Instead, Hooke was taken under the wing of the headmaster of Westminster School, where he was able to study, but also given free rein to continue tinkering with machinery and making things.

Following school, Hooke went to Oxford University, where he became part of a circle of natural philosophers surrounding John Wilkins, the Warden of Wadham College. Wilkins was a polymath who drew together a range of talented individuals including the architect Christopher Wren and the scientist Robert Boyle. Wilkins recognized Hooke's prodigious technical talents, as well as his scientific intelligence and drew him into the group to help further their quest for knowledge. Wilkins inspired Hooke and became a mentor to the young man. It was thanks to Wilkins' influence that Hooke went on to be employed as a technical assistant by Robert Boyle. Hooke was always short of money, so paid work was welcome. Boyle put Hooke to work in 1655 on designing and building an air pump for experiments with a

LEFT: An experiment with an air pump and a glass vacuum sphere.

vacuum in a glass sphere. Hooke's energy and curiosity were boundless and he carried out his own researches alongside this work. In 1660, he formulated the law of elasticity that governs the physics of springs. His equation is known to this day as Hooke's Law. Hooke also improved the mechanism of spring-powered watches by adding a balance spring, an invention that was later credited to Christiaan Huygens (see page 52), much to Hooke's annoyance.

The Royal Society formed in London in 1660 and included some of the key members of the Wilkins group, so it was natural that Hooke, with his multiple talents, would be engaged as Curator of Experiments in 1662. Hooke lectured and demonstrated experiments alongside pursuing his own studies, despite there being a delay in finding a salary for his position. Hooke gradually became a leading figure at the Royal Society. After he received an MA from the University of Oxford in 1663, he was elected a fellow of the Royal Society and with financial security was able to concentrate more on his own work.

Around this time, Hooke built an improved microscope, adding a mirror and a lamp to focus light

on to the subject being examined. The increased clarity aided his studies of everything from micro-organisms to crystals of frozen urine! Hooke found fame when he published *Micrographia*, a book about his discoveries in 1665. Hooke used his artistic talents to produce close-up drawings that revealed everything from the anatomy of a flea and the compound eye of a fly, through to the structure of plant cells. In fact, it is from Hooke's description of cork bark as seen under a microscope that we actually get the word 'cell'. *Micrographia* captured the

ABOVE: Hooke's microscope.

public imagination. The famous diarist Samuel Pepys claimed to have been so entranced that he stayed up all night to read it.

Following the Great Fire of London in 1666, Hooke was appointed Surveyor to the City of London to assist his friend Christopher Wren with rebuilding the city. More than 13,000 houses had burned down, as well as 87 churches and St Paul's Cathedral. Hooke proved very capable in this role, somehow managing to simultaneously continue his work at the Royal Society and as a lecturer at Oxford. However, Hooke found himself increasingly at odds with others, most

BELOW: The flea from *Micrographia.*

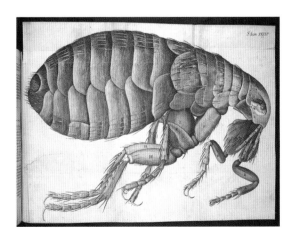

notably with the English scientist Isaac Newton. Both men were jealous of their reputations and clashed over the credit for various discoveries. This would have consequences for Hooke's reputation after he died in 1703. It is thought that Newton never forgave Hooke for the slights on his reputation and when he became president of the Royal Society he suppressed and undermined Hooke's legacy. Thankfully, modern historians have rescued Hooke from being obscured by Newton and we can see how his career marks the beginnings of professional engineering.

THOMAS NEWCOMEN

GREATEST ACHIEVEMENTS

DESIGNED ATMOSPHERIC STEAM ENGINE 1712
This was the world's first effective steam engine. Thousands of Newcomen's design were installed, kick-starting the Industrial Revolution.

ABOVE: A plaque commemorating Thomas Newcomen's steam engine.

'IRON AND HEAT ARE, AS WE KNOW, THE SUPPORTERS, THE BASES, OF THE MECHANIC ARTS. IT IS DOUBTFUL IF THERE BE IN ENGLAND A SINGLE INDUSTRIAL ESTABLISHMENT OF WHICH THE EXISTENCE DOES NOT DEPEND ON THE USE OF THESE AGENTS, AND WHICH DOES NOT FREELY EMPLOY THEM.'

Sadi Carnot, French mechanical engineer, *Reflections on the Motive Power of Fire*, 1824

Engineering made a huge leap forward in Britain in the late 18th century. Inventors and engineers developed new technologies and machinery, which drove a period of rapid transformation that we know as the Industrial Revolution. It overthrew traditional small-scale industries and turned Britain into an industrial powerhouse with mechanized factories and huge cities to house its workers. Engineers also built new transport networks: canals, railways, tunnels and bridges that carried both raw materials and mass-produced products.

The driving force for all this industry and innovation was steam. And the engineer who first successfully harnessed steam power and put it to work, was an ironmonger and metalworker called Thomas Newcomen. In 1712, he invented and built the atmospheric engine: the world's first practical steam engine. It came to be known simply as the Newcomen Engine. More than 2,000 were installed around the world as the Industrial Revolution spread. Newcomen's engine powered the first stages of a dramatic period of change that gave rise to the modern, industrial world.

Thomas Newcomen was born in 1664, in Dartmouth, a town on the estuary of the River Dart in Devon. Nothing is really known about his early career, but he started work as an ironmonger in the south-west of England at around the age of 21. His job would have brought him into regular contact with metalworkers, and Newcomen would have seen at first hand the foundries further north. Immersed in an age of technological innovation, he gradually accumulated engineering expertise and went into partnership with a local plumber, John Calley, who attended the same church. The stage was set for a historic engineering breakthrough by two self-taught engineers that would change the world forever.

Newcomen's customers included the owners of mines in Devon and Cornwall. Demand for metal ores, such as tin, was booming, but mine-owners were unable to fully exploit deposits due to flooding. Coal mines had the same problem. The best engineering solution to drain flooded shafts and galleries at that time was to employ mechanical pumps powered by horses. They hauled a chain of buckets of water from below the ground to the surface. These pumps were both slow and expensive to run. They were also very limited in the amount of water they could remove per day.

BELOW: Newcomen's atmospheric engine.

FAR RIGHT: A replica of the original Newcomen steam engine at the Black Country Living Museum in Dudley.

BELOW: Savery's steam pump.

The need for a more powerful and economical pump was clear. And Newcomen knew there was a new power source available that might help to realize this need: steam. Newcomen had read of Thomas Savery's experiments with a steam pump for draining mines. The principles of Savery's design were sound, using steam to create a vacuum that drew up water. Unfortunately, his pump was very unreliable in practice. It continuously broke down with mechanical faults, or the heat and pressure it operated under caused the air-tight seals needed to maintain a vacuum to fail. However, the main issue with Savery's pump was that it simply could not generate enough suction to be of use in deep mines.

Newcomen and Calley started to experiment with ways to improve Savery's steam pump design. Savery's 'Engine for Raising Water by Fire' had no moving parts other than steam valves. Their major innovation was to introduce a moveable piston housed within a metal cylinder. Initially, they used brass for the cylinder, but later switched to cast iron. Their design condensed steam in this cylinder to create a partial vacuum, which then pulled the piston down. The piston was linked by a chain to a rocking balance beam. This could move up and down like a see-saw, raising and lowering pump rods via another chain. The pump rods drew water into another cylinder, raising it from within the mine and then pumping it out on the surface.

The atmospheric engine proved to be a sturdy and successful design. The first commercially operated pump using an atmospheric engine was installed at a coal mine in Staffordshire in 1712. It was able to pump out just over 45 litres (10 gallons) a minute from a depth of more than 45 m (147 ft). Even better, their design was so robust it could be operated tirelessly day and night to keep mines workable. The atmospheric engine far exceeded the performance of horse-powered pumps. It wasn't long before this new steam-powered technology was rolled out to mines

innovation and despite its wasteful use of fuel (it operated with just 1 per cent energy efficiency) it was used throughout the 18th century and beyond. It provided the basic engineering for steam power to come, including railways and was to be redesigned and perfected by later engineers, most significantly James Watt.

STEAM POWER PIONEERS

There had been earlier attempts to build steam-powered engines. These devices acted as technology demonstrations or prototypes, paving the way for Newcomen.

DENIS PAPIN (1647–1713)

In 1682, the French scientist and engineer Denis Papin demonstrated a pressure cooker, which utilized high-pressure steam and a safety valve. This led Papin to build the world's first steam-driven piston engine in 1690. Papin published his ideas on steam and they may have influenced Newcomen's design.

RIGHT: Denis Papin.

THOMAS SAVERY (1650–1715)

Inspired in part by Papin's work, the British military engineer Thomas Savery built a steam pump for pumping water out of mines. He patented his invention in 1698. Savery's pump didn't have a piston, but used the vacuum created by condensing steam to create suction. His design provided the basic principle of steam power that Newcomen would use in his further development of a steam-powered pump.

LEFT: Thomas Savery.

throughout the country and then exported around the world.

Newcomen's engine boosted production from mines and helped the Industrial Revolution gather pace. The coal and ores and factory-managed goods fundamentally changed the social fabric and economy of Britain and the world beyond. Savery's earlier patent deprived Newcomen and Calley of the full benefit of their invention, but they made a successful living, nevertheless. Between them, the pair saw the installation of more than 100 atmospheric engines throughout Britain and further afield in Europe. The atmospheric engine was a landmark engineering

JOHN
HARRISON

GREATEST ACHIEVEMENTS

GRASSHOPPER ESCAPEMENT 1722

H1 MARINE CHRONOMETER 1735

H4 MARINE CHRONOMETER 1759
This watch set the standard for naval timekeeping and won Harrison the Longitude Prize that he had been working towards for nearly 30 years.

ABOVE: John Harrison with his marine chronometer.

'… IT IS WELL KNOWN BY ALL THAT ARE ACQUAINTED WITH THE ART OF NAVIGATION, THAT NOTHING IS SO MUCH WANTED AND DESIRED AT SEA, AS THE DISCOVERY OF THE LONGITUDE, FOR THE SAFETY AND QUICKNESS OF VOYAGES, THE PRESERVATION OF SHIPS AND THE LIVES OF MEN …'

Excerpt from the 1714 Longitude Act

Engineers find practical solutions to real-world problems. In the 18th century, a major problem for seafaring nations was accurate navigation. The sea was key to global trade and military power, but plotting a course across the ocean could be hazardously imprecise. When ships were out of sight of land, they could use the sun or the stars to determine their latitude, or how far north or south of the equator they were. But there was no accurate way to determine longitude, or how far east or west they had travelled. The British engineer who solved the 'longitude problem' was John Harrison. He revolutionized ocean navigation by developing a super-accurate timepiece that could withstand the rigours of sea travel.

John Harrison was born in 1693, in Foulby, a small village in Yorkshire. His father was a carpenter and John followed him into the trade, before starting a sideline in clockmaking in his early twenties. He

BELOW: The Scilly Isles naval disaster of 1707 was one of the worst maritime accidents in British naval history.

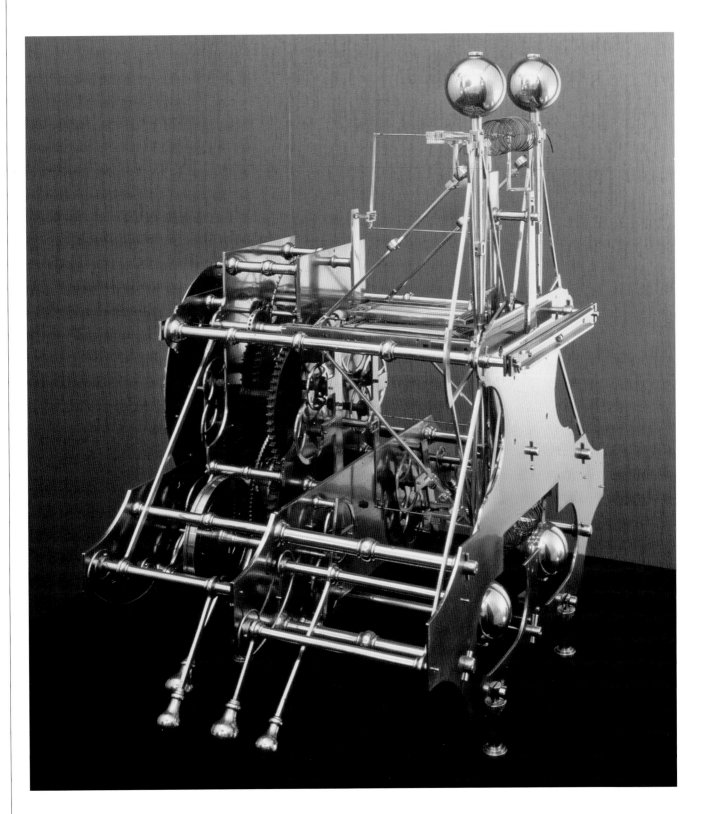

built several pendulum clocks from wood, including their internal mechanisms or 'movements'. The young Harrison was constantly improving these mechanisms, coming up with important innovations to make his clocks ever more accurate. Friction was a major problem, causing clocks to lose time.

Lubrication helped, but had a tendency to gum up the works without continual maintenance. Harrison solved this part of the problem by using an oily, self-lubricating wood called lignum vitae for some of the moving parts. He also developed a new low-friction mechanism to regulate the clock's movements, called a 'grasshopper escapement'. These innovations and the skills in precision engineering he had developed would later stand John Harrison in good stead for developing the marine chronometer.

In October 1707 four British Navy warships ran on to the rocks near the Scilly Isles as a result of inaccurate navigation. They sank and up to 2,000 sailors were lost. This terrible disaster brought what was known as the 'longitude problem' sharply into focus. In July 1714, the Parliament of Great Britain passed the Longitude Act. This offered a prize of £20,000, the equivalent to several million pounds today, to whoever could solve the longitude problem. John Harrison realized the answer lay in building a super-accurate clock. He turned his attention to winning the prize. It would become his life's work.

Harrison realized that if a clock could keep near-perfect time on a sea voyage, it could be used to compare local time with the time back in England. The difference could be used to work out how far east or west you were. At noon, the sun is directly overhead, wherever you are in the world. So a ship's captain could check the ship's clock that had been set at the start of the voyage, and for every hour of difference, 15 degrees of longitude would have been travelled.

Working from his home in Barrow-on-Humber, and with the backing of the Astronomer Royal, Edmund Halley, Harrison set about building his first marine timekeeper, the H1. He knew it would have to be about 50 times more accurate than existing timepieces. It would also have to contend with changes in temperature and the unsteady motion of a ship at sea. Pendulum clocks were the most accurate timepieces on land, but the rolling motion of

a ship affected the travel of the pendulum. Instead, Harrison used a spring-driven movement for the H1, with a balance mechanism that had two swinging spherical weights and a spring instead of a pendulum. His design kept friction to a minimum and the clock didn't require any lubrication. Completed in 1735, the H1 proved accurate during sea trials, but not sufficiently to claim the prize.

Undaunted, Harrison moved to London, where he built two further clocks, the H2 and H3, over the next 19 years. Both involved innovative engineering solutions such as temperature-compensating combinations of metals and more efficient bearings, but Harrison still couldn't claim the prize. Finally, in the 1750s, when Harrison was in his sixties, he changed tack entirely, abandoning a large clock design in favour of a smaller timepiece based on a pocket watch. Harrison's Sea Watch H4 was a masterpiece of precision engineering. The Board of Longitude continued to refuse Harrison the prize, but King George III took up his case and made sure that Harrison was rewarded by Parliament. A later version of the H4 was tested after Harrison's death in 1776 by the famous explorer Captain James Cook, who gave it a glowing review and called it his 'never failing guide'. Measuring just 13 cm (5 in) in diameter, Harrison's small timepiece was a masterpiece of engineering and revolutionized navigation at sea forever.

OPPOSITE: Harrison's precision engineered first marine timekeeper, the H1.

RIGHT: The later and far more portable H4 Sea Watch.

JAMES
WATT

'I CAN THINK OF NOTHING ELSE BUT THIS MACHINE.'

James Watt, letter to Dr Lind, 1765

GREATEST ACHIEVEMENTS

SEPARATE CONDENSER
1764

WATT STEAM ENGINE
1776

SUN AND PLANET GEAR
1781

CENTRIFUGAL GOVERNOR 1788

PORTABLE COPYING MACHINE 1795

ABOVE: James Watt.

Following on from the pioneering work of Thomas Newcomen (see page 58), the Scottish engineer James Watt improved the efficiency and power of steam engines and helped drive the Industrial Revolution to new heights with a machine that could power looms and iron and flour mills.

Born in Greenock on the west coast of Scotland on 19 January 1736, the son of a successful shipwright and shipowner, James Watt suffered ill health in his youth and had to be taught at home. Spending time in his father's workshop, the young James enjoyed building wood and metal models of ships' cranes and capstans, and helped repair nautical instruments. In 1755, following the death of his mother, and as ill health began to afflict his father, Watt moved to London to study as an apprentice scientific instrument maker under John Morgan. Here, he gained experience building rules, scales, quadrants and barometers – tools that demanded the utmost precision.

A prodigy, Watt returned to Scotland within a year and set up his own engineering business in Glasgow. He worked for a while on local canal improvements and the deepening of the Clyde and Forth rivers. His expert repair of delicate astronomical instruments for the University of Glasgow led to the university inviting him to open his own workshop on the campus. Watt began building demonstration models for his university colleague, Joseph Black, the eminent physicist and chemist who discovered carbon dioxide.

In 1764, Watt was tasked with repairing the university's model of a Thomas Newcomen steam engine. Watt realized that Newcomen's design was inefficient. It wasted too much heat and burnt far more coal than was necessary. The engine's cylinder

BELOW: Given a model of Newcomen's steam engine to repair, James Watt instead spent months devising an improvement to the design.

needed to be repeatedly heated to boiling point to produce steam to push a piston, then cooled for the steam to condense and form a vacuum that pulled the piston back. Watt designed a separate chamber for the steam to condense inside, allowing the engine to continue working at one regulated temperature. Watt also introduced lubrication to his upgraded design, reducing friction in the engine's working parts.

Watt spent eight years working as a surveyor and civil engineer to help finance a full-scale working model of his steam-engine design and to purchase a patent for it. He gained the patent in 1769 under the name 'A New Invented Method of Lessening the Consumption of Steam and Fuel in Fire Engines'.

Six years later, Watt formed a partnership with the investor Matthew Boulton to manufacture his steam engines. The Boulton & Watt Company powered the Industrial Revolution with their designs, which could be used anywhere. Over 25 years the pair made their fortunes, building 451 engines, including 268

ABOVE: This working model of a Boulton & Watt engine is based on Watt's double-action design with his patented sun-and-planet gears that help drive a rotating arm.

with rotating arms. At one point a third of the 1,500 steam engines in use around the world originated in Boulton's Soho Manufactory in Birmingham.

The first steam engines were used primarily for pumping water from mines and canals. Watt's design was not only more fuel-efficient – requiring less than one third the amount of coal of Newcomen's design – but it could pump water up from greater depths. The Boulton & Watt engine was in huge demand for operation in Cornish tin mines. Watt began to use the term 'horsepower' as a measure of the work rate of his engines. A typical Watt steam-engine pump could work at a rate equivalent to 56 horses.

The Boulton & Watt company was highly protective of their patent for steam-engine design. Other manufacturers attempting a similar design were taken to court in costly legal battles. While Watt's design continued to be a success, these court actions prevented other engineers from developing potentially better adaptations. Conversely, when Watt attempted to launch an engine with a rotating arm, he had to avoid duplicating a patented design by the inventor James Pickard. Watt's version was described as using a 'sun-and-planet' gear. The 'sun' was a large, toothed wheel, with a smaller 'planet' wheel moving around it, joined to the engine's beam by a swinging rod. The motion of the engine pushed the rod up and down, and caused the planet wheel to orbit the sun.

With the development of a rotating arm, Boulton & Watt engines could produce much more than an up-and-down motion and became practical in paper, iron and flour mills, distilleries, and for powering looms for the weaving trade. Watt continued to improve his designs and invented the double-action engine, which let in steam at both ends of a cylinder, doubling the efficiency of the machine.

While primarily known for his work on steam engines, Watt also devised the first letter copier, to spare himself the effort of making copies of his design plans and drawings. This ingenious device, the forerunner of the modern photocopier, worked by first using a special ink on paper, then pressing dampened tissue paper against it. The tough tissue took a reverse impression of the ink but could be read from the other side.

Watt was made a Fellow of the Royal Society of Edinburgh in 1784 and in London a year later. On his retirement at the dawn of the 19th century, Watt continued to tweak his engine designs as well as fashion musical instruments and a device for duplicating sculptures, from his home workshop at Heathfield Hall, near Birmingham, until his death on 25 August 1819. The workshop and its contents were later relocated to form an exhibit at London's Science Museum. In further recognition of his achievements in engineering, the unit of measurement for electrical and mechanical power – the watt – was named in his honour.

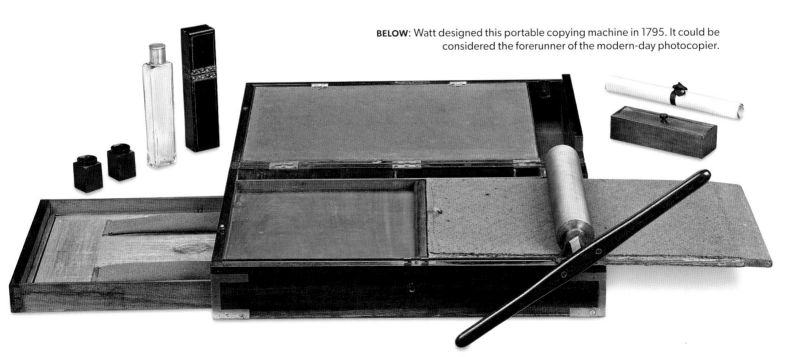

BELOW: Watt designed this portable copying machine in 1795. It could be considered the forerunner of the modern-day photocopier.

THOMAS
TELFORD

'THE WONDER OF IT WAS ALMOST PAST BELIEF...THE MAGICAL STREAM IN THE SKY.'

L. T. C. Rolt on Telford's Pontcysyllte Aqueduct, 1805

GREATEST ACHIEVEMENTS

MONTFORD BRIDGE
Shropshire, 1792

BUILDWAS BRIDGE
Shropshire, 1796

PONTCYSYLLTE AQUEDUCT 1805

MENAI SUSPENSION BRIDGE
Anglesey, Wales, 1826

ABOVE: Thomas Telford.

'The Colossus of Roads', Thomas Telford's skill as a civil engineer changed the map of Scotland. His British road and canal projects, and magnificent bridges are the pinnacle of the early industrial age. The fact that many are still in use underscores his talent for designing exemplary and resilient structures.

Thomas Telford was born on 9 August 1757 at Glendinning, a hill farm in Dumfriesshire, Scotland. His father, a shepherd, died four months later, and Thomas grew up in poverty supported by his mother. Despite this difficult upbringing, young Thomas was a jolly child and known locally as 'Laughing Tam'. Aged 14 he started work as an apprentice stonemason. One of his earliest jobs was to carve a headstone for his father's grave in Westerkirk, Eskdale. His apprentice contribution to a bridge over the River Esk in Langholm, on the Scottish Borders, can still be seen today.

After a period in Edinburgh, Telford moved to London where he worked on additions to Somerset House under the guidance of architects Robert Adam and Sir William Chambers. With most of his training gained through practice, Telford was next given the responsibility of both designing and managing building projects at Portsmouth's Dockyards.

In 1787, Telford became the Surveyor of Public Works in the English county of Shropshire and began a civil building programme in earnest. Among the many projects Telford was responsible for was the renovation of Shrewsbury Castle, the local prison, county infirmary and several churches. One, St Chad's

BELOW: Telford's towering Pontcysyllte Aqueduct over the River Dee opened in Wales in 1805. The aqueduct was designated a UNESCO World Heritage Site in 2009.

Church in Shrewsbury, Telford correctly identified as likely to collapse. It did so just three days later!

Telford gained a reputation as a master bridge designer. His first, built over the River Severn in Montford, was one of 40 he oversaw construction of in Shropshire alone. Inspired by Abraham Darby's famous cast-iron bridge at Ironbridge – an icon of the Industrial Revolution – Telford constructed a bridge in Buildwas, in 1796, that was 9 m (30 ft) wider and half the weight of Darby's. A cautious engineer,

Telford tested the strength of his materials before allowing their use.

In 1799, Telford was called upon to lead a project linking the Ellesmere Canal to the ironworks and coal mines of Wrexham, near the Welsh border, a project that took six years to complete. Telford designed several impressive aqueducts for the canal to bridge rivers and valleys. His Pontcysyllte Aqueduct over the River Dee in Llangollen, Wales, is one of the wonders of the age. Still in use, the aqueduct consists of 19

roads across the Highlands and 1,200 new and improved bridges. He oversaw improvements to several major harbours, including those at Aberdeen and Dundee, the 161 km- (100-mile-) long Caledonian Canal through the remote Highlands, and the building of 32 new churches. Another 296 km of roads across the Scottish Lowlands were completed on his watch.

Telford's expertise was also called upon by the king of Sweden, who employed Telford to build a canal between the capital, Stockholm and Gothenburg city. In 1809 Telford was made a knight of the Swedish royal order in recognition of his services. Continuing his road-building work in Britain, Telford helped rebuild a major trunk road between Holyhead, on the Welsh border, and London. During this period Telford's friend, the poet, and later Telford's biographer, Robert Southey gave Telford the humorous nickname 'Colossus of Roads'.

In 1819, work began on another of Telford's most impressive constructions, the longest suspension bridge in the world at the time, crossing the Menai Strait between North Wales and the island of Anglesey. Supported by 16 522-m- (1,713-ft-) long chain cables made of wrought iron bars, Telford's Menai Suspension Bridge stretched 180 m (590 ft) across the water, with a high clearance for tall ships to pass beneath.

Telford continued to work on road and waterway-improvement projects in his later years, notably on St Katharine Docks, London, the 2.6 km- (1.6 mile-) long Harecastle canal tunnel in Staffordshire, Whitstable harbour, the Birmingham and Liverpool Junction Canal and Galton Bridge in 1829, which boasted the longest single span for a bridge at the time. He was lauded for his work and appointed the first President of the Institution of Civil Engineers, a position he held until his death in London on 2 September 1834. Telford was buried in Westminster Abbey, with a statue erected nearby. In 1968, in respect for Telford's contribution to the county of Shropshire, a new town was named after him.

hollow arches, each 14 m (46 ft) in span, supporting a cast-iron trough 38 m (125 ft) above the valley floor, for a distance of 300 m (984 ft).

Telford found himself sought after to consult on engineering projects across the length and breadth of Britain. He helped advise on improvements to Liverpool's water works and London's docklands. His contribution to the road and canal infrastructure of Scotland is without parallel. Over a period of 20 years Telford led the construction of 1,480 km (920 miles) of

RICHARD
TREVITHICK

GREATEST ACHIEVEMENTS

THE *PUFFING DEVIL*
The first full-size steam locomotive, built in 1801

THE PEN-Y-DARREN LOCOMOTIVE, 1804

CATCH ME WHO CAN 1808

ABOVE: Richard Trevithick.

RIGHT: A replica of Trevithick's wager-winning Pen-y-darren locomotive.

'I HAVE BEEN BRANDED WITH FOLLY AND MADNESS FOR ATTEMPTING WHAT THE WORLD CALLS IMPOSSIBILITIES.'

Life of Richard Trevithick, Francis Trevithick, 1872

Richard Trevithick, the designer of the world's first steam locomotive, made a huge impact on steam power and rail transport but failed to reap the financial rewards and fame of his peers.

Born in Illogan, in the heart of Cornwall's mining lands, on 13 April 1771, Trevithick grew into a tall lad more interested in sport than study. His father, also named Richard, was an engineer at a local copper mine, so his son became familiar with steam-engine technology as it pumped water from the depths of the mine shafts. Unsurprisingly, as soon as he was old enough, the boy left school to work in the mines, quickly gaining promotion to consultant.

Most of the steam engines used in the Cornish mines were stationary water-pumping machines based on Thomas Newcomen's design (see page 58). A much more efficient design by James Watt (see page 66) became popular, but any attempts to duplicate this model were quickly stamped on by the Boulton & Watt lawyers. Trevithick became a local hero when he spoke up in support of engineer Jonathan Hornblower, who had attempted his own modification of Watt's design.

Working at the quaintly named Ding Dong Mine in 1797, Trevithick used his engineering skills to build his own steam engine, one that used high-pressure steam, something that James Watt refused to risk. However, Trevithick then received demands to desist from Boulton & Watt's lawyers. It was not until 1800,

when Watt's patent expired, that Trevithick was free to experiment.

Boiler design had improved over the decades since Newcomen had introduced his steam engine. It was now possible for a skilled engineer like Trevithick to invent a version that used steam at high pressure, without fear of the boiler exploding. This advance meant that steam engines could be designed to do away with Watt's separate condenser. The concept was not new. Indeed, Trevithick's neighbour, the engineer William Murdoch, had tested a model steam carriage with this so-called 'strong steam', but Trevithick was the first to make it a success. Thirty of his more efficient models were built for use in Cornish mines, where they became known as 'puffer whims' for the steam they blew into the air.

High-pressure steam engines were lighter than their predecessors, and safely vented excess steam through a chimney. Trevithick saw the potential to have his steam engine become mobile. He built a full-size steam-powered locomotive in 1801 that could pull a carriage carrying six passengers and demonstrated it to the public in Camborne, Cornwall, on Christmas Eve, towing them uphill to the neighbouring village of Beacon. The vehicle gained the name *Puffing Devil*. Unfortunately, further tests of the locomotive were halted when it broke down three days later. Left unattended, the locomotive boiled away all its water supply, overheated and burnt.

A further setback for Trevithick occurred in 1803, when one of his stationary pumping engines exploded in Greenwich, London, killing four men. Trevithick swore the accident was caused by operator error, but his rivals Boulton & Watt leapt on the chance to raise alarm at the dangers of high-pressure steam. Trevithick moved quickly to introduce a safety mechanism to his design.

An historic moment in steam-locomotive history occurred as the result of a wager. Having commissioned Trevithick to build a steam engine for his Pen-y-darren Ironworks in Merthyr Tydfil, Wales, the owner, Samuel Homfray bet another ironmaster, Richard Crawshay,

LEFT: Trevithick's fourth steam locomotive, the Catch Me Who Can, was demonstrated in 1808 to paying customers.

500 guineas that Trevithick's engine could haul 10 tonnes of iron to the wharf almost 16 km (10 miles) away. Designing a steam locomotive to roll over rail tracks used by horse-drawn carriages, Trevithick proved Homfray right on 21 February 1804, and demonstrated the first steam railway in history.

Four years later, Trevithick put on a show in London with a new locomotive he called *Catch Me Who Can*. The engine chugged around a circular rail track laid near what would become Euston Station. Tickets were a shilling (five pence) each. Disappointingly, the track broke and the locomotive tipped over. Despite several popular demonstrations, Trevithick struggled to convince investors of the merits of his technology. It would be another 17 years until steam-powered locomotives were seriously considered as a replacement for horses.

Facing bankruptcy, Trevithick gave up on locomotives for a while, returning to work on engines for use in mines before an opportunity came his way to work in South America. In the high altitude of the Andes, Boulton & Watt's steam engines proved near useless, while Trevithick's high-pressured machines functioned perfectly. Nine were ordered for Peruvian silver mines. Trevithick left for the country in 1816 aboard a whaler ship but, soon after his arrival, a war of liberation in the country disrupted Trevithick's plans and his opportunities for mining came to nothing. He travelled north to Costa Rica to inspect gold mines before returning through perilous jungle terrain.

Reaching Cartagena in Colombia, defeated and destitute, Trevithick was fortunate to come across the English engineer (and future steam-locomotive builder) Robert Stephenson (see page 84), then working as a silver-mine engineer. Stephenson gave Trevithick £50 to pay for his passage back to England. The Cornish engineer continued to work, improving boiler designs for several years, but with little financial reward. When he died of pneumonia on 22 April 1833, he was penniless and with no friends or relatives nearby to care for him. He was buried in an unmarked grave in Dartford, Kent, but his contribution to steam-engine technology and locomotion would not be forgotten.

GEORGE
CAYLEY

GREATEST
ACHIEVEMENTS

**THE SILVER DISC
MACHINE** 1799
In 1799, Cayley proposed a
fixed-wing aircraft design based
on the scientific principles
of thrust, lift, weight and air
resistance.

CAYLEY'S FIRST GLIDER
1804

WHIRLING ARM 1804

ON AERIAL NAVIGATION
1809–10
A series of papers published in
Nicholson's *Journal* explaining
his discoveries.

**THE GOVERNABLE
PARACHUTE** 1853

ABOVE: George Cayley.

'ABOUT 100 YEARS AGO
AN ENGLISHMAN, SIR
GEORGE CAYLEY, CARRIED
THE SCIENCE OF FLYING
TO A POINT WHICH IT HAD
NEVER REACHED BEFORE
AND WHICH IT SCARCELY
REACHED AGAIN DURING
THE LAST CENTURY.'

Wilbur Wright, aviation pioneer and pilot of the
world's first powered aeroplane, 1909

On 21 November 1783, the Montgolfier brothers' hot-air balloon rose into the skies above Paris with two passengers. The news travelled swiftly around the globe. Finally, the secrets of flight had been mastered. Just over a week later, a hydrogen balloon piloted by another pair of Frenchmen launched from Paris. The exploits of early aeronauts in balloons caught the imagination of the general public. For a time it seemed as if the future of the skies belonged to lighter-than-air aircraft.

News of the Montgolfiers' balloon also reached a nine-year-old English boy called George Cayley, who would one day become a pivotal figure in aviation engineering. However, he would help to steer the future of heavier-than-air aircraft, rather than balloons. Seventy years after the Montgolfiers' flight, when Cayley was an old man, he matched them with his own aeronautical first – building and launching the first glider to carry a person.

George Cayley was the son of a baronet, born in Scarborough, England in 1773. After boarding school, he was taught by private tutors who gave him a solid

ABOVE: A reconstruction
of Cayley's glider, 'The
Governable Parachute'. This
was the first heavier-than-air
aircraft to make a successful
flight.

grounding in mathematics and science, including mechanics and electricity. When his father died in 1792, George became baronet, inheriting the family's estates. This gave him the financial means to pursue his diverse scientific interests, including investigations into the science of flight.

Cayley is remembered as the founding father of aeronautics, but his engineering talents ranged widely. He became a national authority on land drainage and reclamation. He was also a prolific inventor,

who designed self-righting lifeboats, an early type of wire-spoked wheel, a tractor with caterpillar tracks, prosthetic limbs, a gunpowder-fired engine and a hot air expansion engine that both anticipated the internal combustion engine and even invented the first seat belts (for the pilot of one of his gliders).

Cayley established the basic principles of aeronautical engineering. He identified the four forces that an aircraft designer had to contend with to get a heavier-than-air vehicle to fly: lift, weight, thrust

RIGHT: The Montgolfier Brothers' balloon realized the dreams of flight of earlier aviation engineers.

BELOW: A design for a human-powered flying machine by George Cayley – this was one of his ideas that did not get off the ground!

and drag. Cayley succinctly summarized the challenge of aircraft design: 'to make a surface support a given weight by the application of power to the resistance of air'. Theory and practice were literally two sides of the same coin for Cayley. In 1799, he engraved a silver disc with a design of a fixed-wing aircraft on one side and a diagram of air resistance, lift and drag on the other. Cayley had made a decisive break from the widespread idea of using flapping wings to create lift. Even the great Renaissance thinker Leonardo da Vinci had looked to birds as the model for human flight. Instead, Cayley's stationary wing design relied on forward motion to create lift, separating the aircraft's propulsion unit from the winged structure dedicated to lift. His design was unworkable, relying on the muscle power of its pilot to generate lift, but its theoretical basis was sound. Cayley's aircraft only lacked a suitable power plant, a challenge that the aeroplane engineers who followed would overcome.

Without a viable engine, Cayley concentrated on

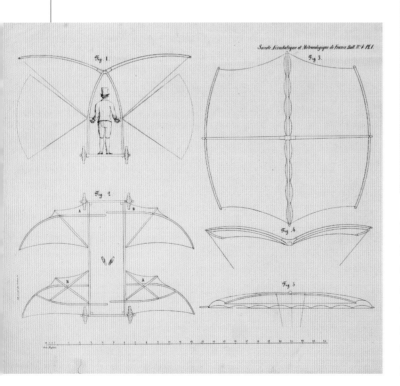

building gliders, using experiments and mathematical models to explore the mechanics of flight. He constructed a Whirling Arm device to test different shapes and angles for wings. With the help of a local mechanic, Thomas Vick, he built and flew numerous model and prototype gliders. Cayley's first gliders were modified kites, but later versions were more recognizably plane-like, with wings, a tail-plane and

a boat-like cockpit below. In 1809 and 1810 Cayley published the knowledge he had gathered from decades of pioneering work in a series of papers in Nicholson's *Journal*. These provided a valuable resource to later aviation engineers.

In 1853, at the age of 80, Cayley designed and built a last glider that he called the Governable Parachute. It was conveyed to nearby Brompton Dale for a test flight. Cayley was too old to fly, so the story goes that the reluctant pilot was his coachman. Following a no-doubt bumpy landing the coachman promptly resigned, despite having made history as the first person to fly in an aircraft that was heavier than air. Replica gliders have been flown in modern times and proved the air-worthiness of Cayley's Governable Parachute, a milestone in aviation engineering.

GEORGE & ROBERT
STEPHENSON

'THE POTENTIALITIES OF STEAM TRANSPORT WERE FULLY REALIZED.'

Thomas Southcliffe Ashton on George Stephenson's Liverpool and Manchester Railway, 1948

GREATEST ACHIEVEMENTS

STOCKTON AND DARLINGTON RAILWAY
1825

LOCOMOTION 1825

ROCKET 1829

LIVERPOOL AND MANCHESTER RAILWAY
1830

CHESTER & HOLYHEAD RAILWAY 1848

HIGH LEVEL BRIDGE
Newcastle, 1849

BRITANNIA BRIDGE
Anglesey, 1850

ABOVE: George Stephenson.

The father and son team of George and Robert Stephenson led the vast expansion of railways across Britain in the 19th century, with George's civil and rail engineering endeavours, and Robert's *Rocket* locomotive inaugurating the passenger-train industry.

The celebrated 'Father of Railways' George Stephenson was born into a mining family in Wylam, Northumberland, England on 9 June 1781. In his youth he worked as a farmhand, a pitboy, then a brakesman, working the winding engine at a local pit. He repaired shoes and clocks to supplement his income. George married a farmer's daughter, Frances Hindmarsh, 13 years his senior, in 1802 and a year later they had a son, Robert (born 16 October 1803). Tragically, three years after Robert's birth, Frances died of tuberculosis.

In 1811 George Stephenson helped improve the water-pumping engines at Killingworth High Pit which led to his employment as an enginewright, responsible for maintaining all of the mine's engines. During this time, Stephenson devised a safety lamp for miners that would burn without causing explosions should the miners encounter a gas leak. This was followed by the design of his first locomotive, *Blücher*, one which was able to move along rails using the friction of its flanged wheels.

On the announcement that a new 40-km- (25-mile-) long railway was to be laid between Stockton and Darlington collieries, Stephenson recommended that, instead of it using horse-drawn coal carts, steam locomotives could move both coal and passengers on

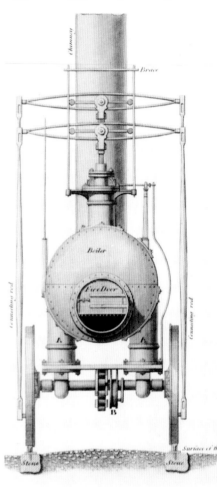

ABOVE: Locomotives for the Stockton and Darlington Railway.

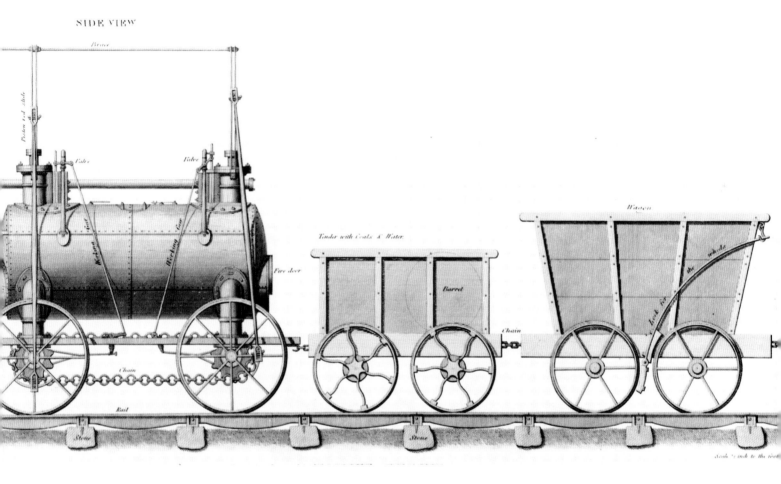

SIDE VIEW

the line. Stephenson surveyed the route in 1821 and set up Robert Stephenson & Co, with his 18-year-old son as a managing partner, to manufacture the locomotives. The Stockton and Darlington Railway opened on 27 September 1825 with George Stephenson himself driving the engine *Locomotion,* as it pulled wagons loaded with 80 tonnes of coal and flour, along the newly laid wrought-iron tracks. Stephenson chose a gauge of 4ft 8½ in (1.5 m) for the tracks, which became the standard for railways over much of the world. A purpose-built passenger car was attached to *Locomotion* to convey VIPs for the inaugural journey. About 600 passengers hopped aboard for the maiden journey – the first to travel on a steam-locomotive railway.

A year later, George was leading the construction of the Liverpool and Manchester Railway, a huge civil-engineering project which linked the major English cities via 50 km (31 miles) of railway. The route crossed dangerous peat bogs but Stephenson overcame this problem by supporting the tracks on wood and shingle.

While work was going on with the track laying, a competition was held to design a locomotive to operate on the line. All the entrants for 1829's Rainhill Trials had to be lighter than 6 tonnes and travel back and forth at an average speed of 16 km/h (10 mph) for 56 km (35 miles). Of the five that ran, only one steam locomotive succeeded – *Rocket*, designed by Robert Stephenson, with help from his father.

The Liverpool and Manchester Railway took four years to complete. The Prime Minister, the Duke of Wellington, was one of many dignitaries who turned

out for the opening ceremony, which was sadly marred by the death of the Liverpool MP William Huskisson, who was struck by *Rocket* as he was crossing the tracks.

George Stephenson continued to advise on railway expansion throughout the rest of his life, and was much in demand. Under his and Robert's guidance new railway routes were built in Leicestershire, between Derby and Leeds, Normanton and York, Manchester and Leeds, Birmingham and Derby, and Sheffield and Rotherham. It was the peak of Victorian railway expansion, with the Stephensons, along with George Stephenson's former apprentice, Joseph Locke and Isambard Kingdom Brunel (see page 92), at the forefront. Railroad builders sailed from the United States to investigate George Stephenson's work. Subsequently, Stephensons' locomotives were exported for use in North America.

Both George and Robert Stephenson were offered knighthoods for their contributions to Britain's rail network but they declined. In George's words, he wanted 'no flourishes' to his name. Though, in his

retirement, in 1847, George Stephenson did accept the appointment as the first president of the Institution of Mechanical Engineers. He died a year later on 12 August 1848 in Chesterfield, Derbyshire.

During his father's final decade Robert Stephenson was becoming equally famous for his rail infrastructure that included several ground-breaking bridge constructions. One of his rail routes, the Newcastle to Berwick line, required the building of 110 bridges, including the impressive High Level Bridge at Newcastle.

For the Chester & Holyhead line, Robert was tasked with designing a rail bridge to cross the Menai Strait between the North Wales mainland and the island of Anglesey, within sight of Thomas Telford's (see page 72) record-breaking suspension bridge. He approached the task with trepidation following the collapse of one of his railway bridges near Chester, that had resulted in the deaths of five people. After months of testing materials, alongside fellow engineer William Fairbairn, work began on the Britannia bridge using rectangular tubes of wrought iron for trains to pass through. The completed bridge was a success and remained in use as it was for 120 years. Stephenson's revolutionary engineering work led to further constructions under his guidance as far afield as Canada and Egypt.

Robert Stephenson married in 1829 but never had children. His wife died in 1842, aged just 39 and Robert never remarried. Suffering from poor health throughout most of his life, Stephenson was forced to convalesce in 1859. He chose to take time off on a yachting break, but had to end it early when his health declined further. He died at his home in London on 12 October 1859. Crowds turned out to watch his coffin pass on its way to Westminster Abbey.

Both Robert and his father left behind a huge legacy, with their locomotive designs leading to further engineering advancements. Many of their bridges and track routes are still crossed by trains today.

MICHAEL FARADAY

'NOTHING IS TOO WONDERFUL TO BE TRUE, IF IT BE CONSISTENT WITH THE LAWS OF NATURE, AND IN SUCH THINGS AS THESE, EXPERIMENT IS THE BEST TEST OF SUCH CONSISTENCY.'

Michael Faraday, Diary, 1849

GREATEST ACHIEVEMENTS

ELECTROMAGNETIC ROTATION 1821

ANNUAL CHRISTMAS LECTURE AT THE ROYAL INSTITUTION
Began 1825

DISCOVERY OF ELECTROMAGNETIC INDUCTION 1831

FARADAY DISK
The first electric generator, 1831

LIGHTHOUSE CHIMNEY
1843
The only invention of Faraday's to be patented.

CARBON ARC LAMP IN SOUTH FORELAND LIGHTHOUSE 1858

ABOVE: Michael Faraday.

Michael Faraday's experiments with electricity led to the design of the electric motor and modern electrical engineering, while his annual lectures inspired a new generation of scientists and engineers.

When Michael Faraday was born on 22 September 1791, electricity was something of a novelty with little expectation of it having a practical use. Faraday grew up in Newington Butts, London, the third of four children with his father barely able to work as a blacksmith. At the age of 14 he was sent to work as an apprentice bookbinder under George Riebau in the Bloomsbury area of London. Here, he became fascinated with the contents as much as the fabric of the books in his hands, particularly those about science. With Riebau's blessing, the young Faraday took over a back room in the bookshop as his own laboratory.

Invited by a customer, William Dance, Faraday began attending Royal Institution lectures delivered by Humphry Davy, a chemist renowned for discovering the properties of nitrous oxide (laughing gas). Faraday kept notes of the lectures and bound them for Davy, hoping his gift might lead to him

OPPOSITE: As director of the Royal Institution, Faraday established the tradition of the Christmas lecture, bringing scientific enlightenment to generations of children.

BELOW: In one of his most productive experiments Faraday wrapped two wires around an iron ring. Sending an electric current through one wire induced a magnetic field in the ring which then induced an electric current in the second wire.

being taken on as his assistant. When the position became free a year later, in 1813, Davy remembered Faraday and invited him to join his tour of continental Europe. The pair visited laboratories and eminent scientists of the day, such as André-Marie Ampère and Alessandro Volta, an invaluable apprenticeship for the budding science student, Faraday.

In 1820, the Danish scientist Hans Christian Ørsted set the scientific world alight with his discovery that a magnetic compass needle moved when in close contact with an electric current. This new phenomenon was called electro-magnetism. Engaged by Davy to investigate the findings, Faraday observed, on 4 September 1821, that a vertical wire, dipped into a pool of mercury, and carrying an electrical current, would rotate around a magnet. As a result, it would be possible to generate continuous motion using electricity and magnetism, the concept behind the electric motor.

When Faraday failed to acknowledge Davy's contribution to his discovery, Davy had his assistant undertake a mostly pointless six-year study of optical glass for scientific instruments. It was another ten years before Faraday made his next major breakthrough, the discovery of electromagnetic induction, something recognized independently by the US scientist Joseph Henry. Faraday proved it was possible to generate an electric current through the movement of magnetic fields, which Faraday described as 'lines of force'. Faraday's experiments led to his effective invention of the transformer and dynamo.

Another area in which Faraday made a huge contribution was lighthouse technology. He instigated the use of rotating lights at a fixed speed which sailors could use to gauge their position. He also tested and introduced electric light to two lighthouses. Though the systems did not become permanent, as part of Faraday's effort he proved the usefulness of one of his inventions, the electricity generator.

Following on from his mentor, Davy, Faraday began lecturing regularly at the Royal Institution. On his promotion to director, Faraday established the annual Christmas lectures for children in 1825, and led 19 of them. The most famous of these lectures was 'The Chemical History of a Candle' in which he used a simple candle as a launchpad into an explanation of multiple scientific ideas. The notes from this lecture have been in constant publication ever since, while the Royal Institution's tradition of entertaining talks on popular science has continued to the present day.

In his sixties Faraday was forced to reduce his workload. He had begun to suffer regular headaches, spells of dizziness and memory loss, which made it difficult for him to write anything with clarity. Faraday saw out his last days in an apartment at Hampton Court Palace, provided for him by Queen Victoria. He died on 25 August 1867, but he kick-started a revolution in the generation of electricity. As a result of Faraday's discoveries, the world would be transformed as electricity powered a new modern age of technology.

ISAMBARD KINGDOM
BRUNEL

'THE CHARACTERISTIC FEATURE OF HIS WORKS WAS THEIR SIZE, AND HIS BESETTING FAULT WAS A SEEKING FOR NOVELTY...'

Obituary, Minutes of the Proceedings of the Institution of Civil Engineers, 1860

GREATEST ACHIEVEMENTS

MAIDENHEAD BRIDGE
Completed in 1839

BOX TUNNEL
Box Hill, Wiltshire 1841

BRISTOL TEMPLE MEADS STATION 1841

GREAT WESTERN RAILWAY
Route from London to Bristol 1841

THAMES TUNNEL 1843

SS *GREAT EASTERN* 1858

ROYAL ALBERT BRIDGE 1859

CLIFTON SUSPENSION BRIDGE
Completed in 1864 as a tribute to Brunel by the Institute of Civil Engineers.

ABOVE: Isambard Kingdom Brunel.

RIGHT: The Thames Tunnel was an engineering marvel and a major London tourist attraction while it was being built. Fifty thousand people walked through the tunnel on the day it opened.

Isambard Kingdom Brunel was one of the foremost engineers of the Victorian era, a restless innovator whose genius, drive and determination propelled the world towards modernity. Brunel brought new approaches and tireless energy to the building of bridges, tunnels, docks, viaducts, railways and steamships. He is chiefly remembered as a pioneer of the railways that revolutionized transport networks. The railways changed everyday life in the 19th century, with an impact that can only be compared with that of the invention of the internet on the modern world.

Isambard Kingdom Brunel was born in Portsmouth, England, on 9 April 1806. He was the youngest child of French émigré Marc Brunel, an engineer and inventor who prepared his son for a career in engineering from an early age. The young Brunel was taught to draw from the age of four and then received the best education his father could afford. He was sent to school in France to learn the most advanced mathematics of the day. Having completed his studies and following a short

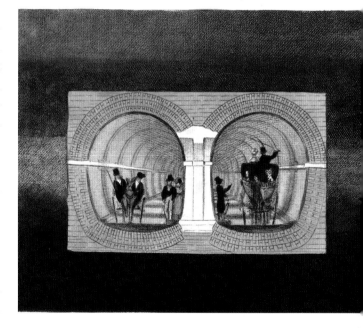

apprenticeship with a leading clockmaker, Isambard returned to Britain in 1822 to start work.

Isambard was barely 20 years old when he joined his father as assistant engineer on a project to build a tunnel under the River Thames in London. Marc Brunel had invented a protective tunnelling shield to help excavate a tunnel through the unstable riverbed sediments. But even with the shield the Thames Tunnel remained a risky venture. The tunnel was flooded twice and on the second occasion six men were drowned and an unconscious Isambard was the only survivor pulled from the water. Isambard eventually took on the project management of the Thames Tunnel from his father. When it opened in 1843 it was the world's first underwater tunnel.

While convalescing in Bristol after his accident in the Thames Tunnel, a restless Isambard entered a competition to build a bridge across the Avon Gorge. His design for the Clifton Suspension Bridge won, despite opposition by engineering rival Thomas Telford (see page 72), who was the judge! Sadly, due to a lack of finance, the bridge wouldn't be built

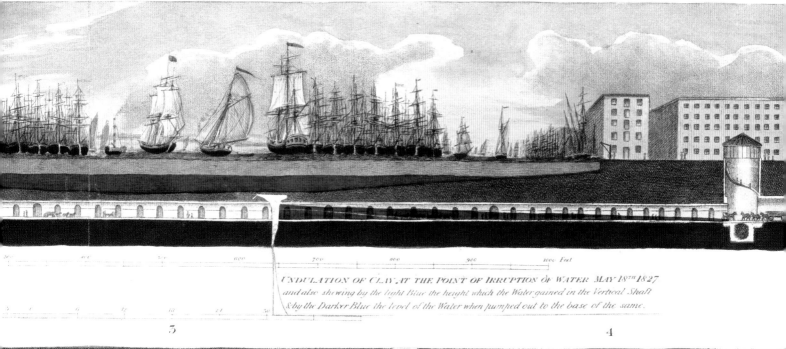

UNDULATION OF CLAY AT THE POINT OF IRRUPTION OF WATER MAY 18TH 1827
and also shewing by the light Blue the height which the Water gained in the Vertical Shaft
& by the Darker Blue the level of the Water when pumped out to the base of the same.

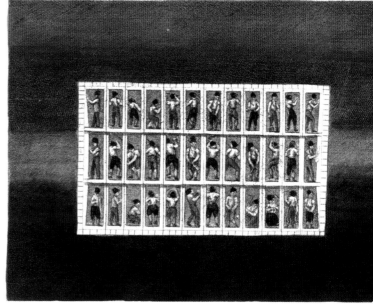

until after Brunel's death. But the ambitious young engineer had proved his talent with his first major independent design, and his stay in Bristol led to him rebuilding the docks there, the first of several dock projects.

Brunel's career took off in 1833, when he was appointed chief engineer of the Great Western Railway. He was just 27 years old when he took charge of the most ambitious railway building project the world had seen to date. Isambard personally surveyed the route, travelling and sleeping in a horse-drawn carriage that served as his portable office. To construct

RIGHT: As part of his Great Western Railway, Brunel built the Royal Albert Bridge with its unique tubular design across the Tamar River to link Devon and Cornwall.

LEFT: Including the Great Western Railway, Brunel built more than 1,900 km (1,180 miles) of railways in Britain, Ireland, Italy and East Bengal.

94

the railway line with as minimal a gradient as possible, Brunel built numerous bridges, tunnels and viaducts. Some of these structures were ground-breaking works of engineering. His bridge over the Thames at Maidenhead had two of the flattest brick arches ever built. Box Tunnel was nearly 3 km (2 miles) long – longer than any tunnel built before. It took two shifts of 1,500 labourers working by candlelight day and night for five years to complete. Working from either end, the accuracy of Brunel's calculations ensured that the two tunnels driven through solid rock joined up almost perfectly.

A controversial aspect of Brunel's railways was the broad gauge he adopted for its tracks. His rails were set just over 2 m (6 ft 6 in) apart so as to provide a smoother ride than Stephenson's already established narrower gauge railways (see page 84). The competition between the two track standards that competed to expand across Britain became known as the 'Gauge Wars'. Brunel's broad-gauge standard eventually lost, but his achievement in building the Great Western Railway was a triumph. It was also a testimony to Brunel's determination and capacity for hard work. In addition to solving the railway's engineering challenges, Brunel also had to deal with social and political obstacles. He had to assuage public fears about high-speed railways, as well as overcoming resistance from vested interests such as the stagecoach and canal companies, and landowners who didn't want railways driven across their estates.

Having dominated railways in the west of Britain, Brunel sought to extend their reach across the Atlantic with steamships. His idea was that a passenger could board a GWR train in London and travel to New York in America. With this integrated transport network in mind, he built some of the most revolutionary ships in history. The SS *Great Western* was the first, a giant paddle-wheel steamship that was the world's largest passenger ship on launch in 1838. The SS *Great Britain* followed in 1843, the world's first ocean-going steamship with an iron hull and powered by a screw propeller.

Brunel topped his first two giant ships with the SS *Great Eastern* in 1858, the largest ship ever built at that time. It was designed to carry 4,000 passengers to Australia without refuelling. However, due to spiralling development costs, technical failures and accidents, neither the SS *Great Britain* nor the SS *Great Eastern* ever fulfilled their potential. The building of the gigantic SS *Great Eastern* proved to be Brunel's last major project. Shortly after its maiden voyage he suffered a stroke and died ten days later on 15 September 1859. He was 53 years old.

Brunel was an engineer who thought and built on a grand scale. The size of his projects reflected the scope of his ambitions, from vast railways and their infrastructure to giant ships larger than any constructed before. His career had its share of failures, both technical and financial. The experimental atmospheric railway Brunel opened in 1847 was abandoned after a year of technical problems, while his steamship companies often teetered on the edge of bankruptcy. But setbacks were to be expected for an engineer who worked at the cutting edge of what was possible. Today, Brunel is rightly seen as the most iconic engineer of the steam age, with a leading role in the technological revolution that led to the modern world.

RIGHT: Though it was a commercial failure as a passenger liner, Brunel's colossal SS *Great Eastern* was later used to lay the world's first transatlantic telegraph cable.

JOHN, WASHINGTON & EMILY ROEBLING

GREATEST ACHIEVEMENTS

BROOKLYN BRIDGE
New York, 1883

ABOVE: John Augustus Roebling

'THE BUILDERS OF THE BRIDGE. DEDICATED TO THE MEMORY OF EMILY WARREN ROEBLING WHOSE FAITH AND COURAGE HELPED HER STRICKEN HUSBAND COLONEL WASHINGTON A. ROEBLING COMPLETE THE CONSTRUCTION OF THIS BRIDGE FROM THE PLANS OF HIS FATHER JOHN A. ROEBLING WHO GAVE HIS LIFE TO THE BRIDGE.'

Plaque on Brooklyn Bridge erected by Brooklyn Engineers Club, 1931

On 24 May 1883, a huge crowd watched as the president of the United States and the governor of New York led celebrations for the opening of the Brooklyn Bridge. Thousands had travelled from far and wide to see this marvel of modern engineering. Within 24 hours of its opening, an estimated 250,000 of them had crossed the East River via the bridge's raised promenade walkway. More than a century later, this landmark bridge is still going strong, carrying more than 100,000 cars and thousands of pedestrians every day.

John Augustus Roebling was the civil engineer who designed the Brooklyn Bridge. Born in Prussia, he had studied architecture, engineering and bridge building in Berlin, be-

fore emigrating to America. After a spell as a farmer, Roebling had returned to surveying and engineering. While working on a railway designed to haul canal boats, he spotted the potential for using ropes made of durable metal wire rather than traditional hemp. Soon Roebling had devised a process to manufacture ropes out of wrought iron wires. From there he

RIGHT: The construction of Brooklyn Bridge.

progressed to designing bridges and aqueducts using wire-rope suspension. Having built several major suspension bridges, Roebling was well-qualified to take on the Brooklyn Bridge project as chief engineer.

Roebling came up with a pioneering hybrid design that combined elements of a cable-stayed bridge and a suspension bridge. The Brooklyn Bridge would be the world's first suspension bridge to use steel wires, with the strength made it possible to build on a grand scale. Sadly, a fatal accident meant that Roebling did not live to see his visionary bridge built. While surveying the site, his foot was crushed by an incoming ferry. Refusing conventional medical treatment, tetanus set in and just three weeks later Roebling died.

John Roebling's eldest son, Washington, had been working as his assistant and was now appointed chief engineer of the Brooklyn Bridge. He was already an accomplished engineer in his own right. Washington had studied engineering in New York and during the Civil War had built suspension bridges for the Union Army. After the war, following his marriage to Emily Warren, Washington worked with Roebling Senior on a number of large suspension bridge projects. He had travelled to Europe in the preceding years to study pneumatic caissons. These were large open-bottomed chambers, pumped full of compressed air that could be used to build bridge foundations underwater.

In 1870, work began on the suspension tower on the Brooklyn side. The first of two caissons was lowered to the riverbed and teams of labourers started to excavate through the mud to lay the foundations. As the men dug deeper, the pressure got worse. Workers started to suffer from 'caisson disease'. Today, divers know it as 'the bends', a dangerous condition caused by returning too quickly to the surface. Several men died and Washington collapsed and nearly died from caisson disease in 1872.

Washington survived, but his health was shattered. Largely housebound from then on, he continued as chief engineer by sending a steady flow of detailed instructions to his assistants, who then sent reports back to him. Emily shuttled between the house and the bridge with messages. Soon she was deeply involved with the technical, logistical and political problems of the bridge. As the build progressed, the Roeblings moved closer to the bridge and Washington could survey progress from the house. Slowly, but steadily, two 85-m- (279-ft-) high suspension towers of granite and limestone grew to dominate the skyline over the river. Four thick steel cables were strung across the towers and anchored onshore. Then steel wires were dropped from the main cables to suspend the deck that would carry

traffic across the bridge. Finally, after more than 13 years of intense work, the bridge was complete.

On opening day in 1883, Emily Roebling had pride of place in the first carriage to cross Brooklyn Bridge. She carried a rooster as a symbol of victory. It had been a hard-won victory, but together the Roeblings had finally realized the dream of a bridge that spanned the East River. Measuring 1,825 m (5,987 ft) end-to-end, the Brooklyn Bridge was the largest suspension bridge of its day, an engineering masterpiece that was acclaimed as the eighth wonder of the world.

RIGHT: New York's Brooklyn Bridge was designated a Historic Civil Engineering Landmark by the American Society of Civil Engineers in 1972.

BELOW: Emily Roebling.

JOSEPH BAZALGETTE

'HE PUT THE RIVER IN CHAINS.'

Translation from Latin on Joseph Bazalgette memorial, London, 1901

GREATEST ACHIEVEMENTS

DEPTFORD PUMPING STATION 1864

LONDON SEWER SYSTEM 1865

ALBERT EMBANKMENT 1869

VICTORIA EMBANKMENT 1870

ALBERT BRIDGE 1884

PUTNEY BRIDGE 1886

HAMMERSMITH BRIDGE 1887

BATTERSEA BRIDGE 1890

ABOVE: Joseph Bazalgette.

RIGHT: The reservoir built at Crossness was large enough to contain 120 million litres (26 million gallons) of sewage before releasing it at ebb tide into the Thames estuary.

With his major reconstruction of London's sewers in the early 19th century, Joseph Bazalgette was responsible for saving thousands of lives from the peril of cholera, while transforming the layout of the capital city.

Joseph Bazalgette was born to parents of French descent on 28 March 1819 in Enfield, North London. His father, also called Joseph, was a Royal Navy captain who had been wounded in the Napoleonic Wars. Little is known of young Joseph's early years except that, in 1835, he began his distinguished engineering career in Northern Ireland. Under the guidance of Irish civil engineer Sir John Macneill, Bazalgette helped with land-drainage work, something that would stand him in good stead for his future undertakings. Seven years later, Bazalgette had taken on the role of consulting engineer for Westminster, London. The pressures of overwork on the railways affected Bazalgette's health, however, and he had to quit.

In the first half of the 19th century, the population of London had almost doubled, putting a strain on the antiquated sewage system. Most human waste found its way into the River Thames, leaving it as one of the world's most polluted waterways. Despite its

filth, the Thames was used for cleaning and even drinking water for much of the population. The effect on public health was devastating, with water-borne contagions, such as cholera, killing thousands. Infant mortality rose to 20 per cent and life expectancy fell to below 30 years. More than 14,000 Londoners died due to a cholera outbreak between 1848 and 1849.

In 1858, the 'Great Stink' issued from the Thames and, at the Palace of Westminster, curtains had to be drawn and soaked in lime chloride to cover the stench. The British government was compelled to act.

Bazalgette was now chief engineer for the London Metropolitan Board of Works and was provided with the vast funds needed to update London's ailing sewers.

Bazalgette's plan was to build a new underground system that would redirect effluent downstream from the city to Erith Marshes in the Thames Estuary. The network required four new pumping stations, 2,100 km (1,305 miles) of sewers and 131 km (81 miles) of

FLVM'N' · VINC'LA · P°SV'T

Sir JOSEPH BAZALGETTE CB
Engineer of the London Main Drainage System
and of this Embankment

intercepting sewers joining from the east and west, with 320 million bricks used in the construction. On arrival at Erith Marshes, the sewage would be stored in tanks until its release into the Thames at ebb tide, when it would be flushed far away from the bustling capital. The system was officially opened by Edward, Prince of Wales in 1865, ten years before its completion.

At the time, it was thought that cholera was transmitted through the air as a miasma and that flushing waste below ground would prevent the foul air from affecting people. Instead, it was the avoidance of polluted waters that helped prevent further outbreaks.

Bazalgette had the foresight to make the sewer pipes twice the diameter recommended, assuming the population of London would continue to increase. He was, of course, correct, and his sewage system was able to cope with the capital's needs for more than a hundred years.

Bazalgette not only controlled what passed beneath London's streets, he also remapped above ground. As a measure for restricting the river's flow, Bazalgette ordered the reclamation of land on both Thames riversides, creating the new embankments Albert, Chelsea and Victoria. The 2 km- (1.25 mile-) long Victoria Embankment near Westminster incorporated a subway and an underground railway. The work also helped traffic flow in the city and provided an estimated an extra 21 hectares (52 acres) of land. Bazalgette also designed three new bridges over the Thames, at Battersea, Hammersmith and Putney, and oversaw plans for 3,000 new streets. Few engineers could take as much credit as Bazalgette for transforming the map and well-being of a city.

Joseph Bazalgette was knighted in 1874. Following in the footsteps of both Thomas Telford (see page 72) and George Stephenson (see page 84), Bazalgette was made president of the Institution of Civil Engineers in 1884. By the time of his death on 15 March 1891, cholera was no longer feared by Londoners, who could breathe deep without discomfort.

RIGHT: Bazalgette designed three new embankments in London, reclaiming land beside the Thames. The Victoria Embankment incorporated two new public gardens.

NIKOLAUS
OTTO

'HIS CAREER EXEMPLIFIES THE SUCCESS OF PERSEVERANCE AND ENERGY PAIRED WITH SKILL AND INGENUITY.'

Obituary for Nikolaus Otto, *Engineering*, 1891

GREATEST ACHIEVEMENTS

FIRST PETROL-POWERED ENGINE 1861

FIRST EXPERIMENTS WITH FOUR-STROKE ENGINES 1862

THE ATMOSPHERIC ENGINE 1864

ESTABLISHES THE FIRST ENGINE FACTORY IN THE WORLD 1864

GOLD MEDAL IN WORLD EXHIBITION AT PARIS FOR ATMOSPHERIC ENGINE 1867

DEVELOPMENT OF FOUR-STROKE ENGINE (OTTO-MOTOR) 1876

INVENTS THE ELECTRIC IGNITION 1884

ABOVE: Nikolaus Otto.

Though he had no technical training, Nikolaus August Otto successfully designed an internal combustion engine that ran on liquid fuel. His design replaced the steam engine as the powerhouse of industry, selling tens of thousands.

Nikolaus August Otto was born on 10 June 1832 in Holzhausen an der Heide, a German village beside the River Rhine. His father, a postmaster, died within months of Nikolaus' birth. Under the care of his mother the boy did well at school and seemed

BELOW: Nikolaus Otto built the first liquid-fuelled four-stroke engine. More than 50,000 such engines were sold in his lifetime.

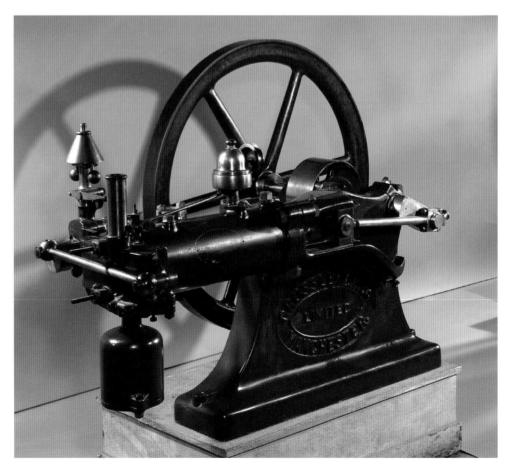

set for a technical education but, with the German economy suffering, he was pushed towards a career as a merchant.

On leaving school, Otto worked as a grocery store clerk before moving to Frankfurt, then Cologne, to get a job in sales, providing imported tea, sugar and kitchenware for stores in western Germany. While travelling as a sales representative, Otto gained an interest in engines. Across the border, in France, the engineer Étienne Lenoir had just developed the first working internal combustion engine. It worked like a steam engine, but with a piston powered by the explosive ignition of air and coal gas. While Lenoir's design worked, it was not very practical. It gave off a lot of heat, and fuel had to be delivered as a gas. It proved expensive to run. Otto began conceiving of an improved internal combustion engine, but one that worked using liquid fuel.

Though lacking in a technical education, Otto built the first petrol-powered engine in 1861. Forming a partnership with the manufacturer Eugen Langen in 1864, Otto was able to develop his design and present it three years later at the Paris World Exhibition of 1867. The Otto-Langen engine impressed a panel of judges with its efficiency and was awarded a gold medal.

This result led to huge demand for Otto's internal combustion engine, but his company struggled to satisfy. Despite another partnership, with Ludwig August Roosen-Runge, a businessman from Hamburg, not enough engines could be produced. Finally, Langen managed to convince further investors, including his own brothers, to back the project. A new company, Gasmotoren-Fabrik Deutz AG, began trading in January 1872. Among the staff Langen hired for the company were Gottlieb Daimler (see page 114) and Wilhelm Maybach.

Deutz AG took off to be the number-one engine manufacturer in the world as a result of Otto's superlative design, which he improved to incorporate four strokes. The concept of the four-stroke engine had been patented in 1862 by a French engineer, Alphonse Beau de Rochas, but Otto was the first to build one. In a four-stroke engine, the cycle of combustion took place over four complete strokes of the piston. When the piston moved up, the air-fuel mix was drawn in. On the second stroke, the mix was compressed. It was lit on the third and, on the fourth, the piston pushed the exhaust gases out. More than 30,000 of these so-called Otto engines were sold within ten years.

Otto found it difficult to retain the patent on his design once the work of Beau de Rochas was uncovered, and by 1889 more than 50 companies were manufacturing their own four-stroke engines. Otto's engines were lighter than steam engines and required less manpower to keep them running but their designer showed little interest in adapting them for use in transportation. Instead, most were installed in factories. It needed another pair of bright sparks to take the initiative. In 1889, Daimler and Maybach, who had left Deutz AG seven years earlier, fitted a four-stroke engine into a carriage to create the world's first four-wheeled automotive. A year later, the first Daimler motor vehicles went on sale.

Otto died in Cologne on 26 January 1891. Despite the competition, Otto died a wealthy man and left behind an invention that would drive the automobile industry.

WILLIAM LE BARON

JENNEY

'THE TRUE FATHER OF THE SKYSCRAPER.'

Committee report on Jenney's Home Insurance Building, 1931

GREATEST ACHIEVEMENTS

FIRST LEITER BUILDING
Chicago 1878

HOME INSURANCE BUILDING
Chicago 1885

MANHATTAN BUILDING
Chicago 1891

LUDINGTON BUILDING
Chicago, built in 1891, National Historic Landmark

NEW YORK LIFE INSURANCE BUILDING
Chicago 1894

HORTICULTURAL BUILDING
For the World's Columbian Exposition, Chicago 1893

ABOVE: William Le Baron Jenney.

Engineer and architect William Le Baron Jenney raised the Chicago city skyline to new heights with his mastery over materials and design, building the world's first 'skyscraper'.

Jenney was born on 25 September 1832, in Fairhaven, a small town in Massachusetts, USA. His father was the successful owner of a fleet of whaling ships. Thanks to his family's wealth, Jenney benefited from a good education, studying at Phillips Academy in Andover before moving to the west coast to join the Californian gold rush of 1849 while still in his teens.

One year later, San Francisco was devastated by a great fire that destroyed most of its wooden buildings. Jenney was witness to its rebirth with more resilient brick structures. His fascination with architecture and engineering was further kindled by travels in the Philippines and the South Sea. Here, he admired native buildings constructed around light, flexible bamboo frames which could withstand tropical storms.

By 1851 Jenney was set on a career in engineering. He returned to the States to enrol at Harvard University but, dissatisfied by the quality of his course, left to train in France, home to a wave of modern-thinking engineers. At the École Centrale des Arts et Manufactures in Paris, Jenney studied the work of the influential structural designer Jean-Nicolas-Louis Durand and became familiar with the latest methods of construction with iron. He graduated one year after his classmate Gustave Eiffel (see box on page 113).

Once qualified, Jenney used his training abroad, working as an engineer for a Mexican railroad company, then designing a mechanical bakery for the French army. In 1861, the American Civil War broke out and Jenney interrupted his career to enlist as an engineering officer under General Ulysses S Grant. During his six years in the Union Army, he progressed to the post of chief engineer, designing fortifications and gained the rank of major, which he proudly held for the rest of his life.

After the war, Jenney married and moved to Chicago to open his own architectural office in 1868. Two years later he won a contract to design the West Chicago Park System, breathing life into the city, with three new parks, wide boulevards, lawns and water features.

Just as had occurred in San Francisco, Chicago was razed by fire in 1871 – with 9 km^2 (3.5 square miles) of the fourth largest city in the US turned to ash. Jenney's firm was much in need to help with the rebuilding and Jenney soon gained a reputation for his innovative office buildings. Instead of wood and bricks, Jenney used iron columns as a framework for construction, starting with the First Leiter Building in 1878. Made of lighter materials, and not relying on a supporting wall, Jenney's buildings could incorporate more windows and extra floors could be added.

OPPOSITE: The Home Insurance Building in Chicago was described as 'the most important building of its era, and one of the most important in American history'. Sadly, this pioneering building was demolished in 1931.

THE CHICAGO BUILDING OF   THE HOME INSURANCE CO.

OF NEW YORK

LEFT: Built in 1891, the Manhattan Building was the first 16-storey commercial building in the United States. It is now the oldest surviving skyscraper to use a metal skeletal supporting structure.

The Leiter Building was forward-looking for another reason. It incorporated Elisha Graves Otis' recent invention, the elevator with safety brake, for travel between floors.

Jenney's Home Insurance Building, opened in 1885, was another pioneering work, that laid the foundations for building skyscrapers. It was the first building in the United States to be built around a fireproof steel and iron frame of columns and beams, inside and outside. Though unimpressive by today's standards, at 11 storeys high, the Home Insurance Building was regarded as the first skyscraper. The only way was up, as Jenney's business thrived. He continued using iron and steel in his designs with Chicago's Manhattan Building, completed in 1891, being the first 16-storey commercial building in the US.

Aged 73, Jenney had to retire on health grounds. His final project, which he was unable to complete, was the Vicksburg battlefield memorial in Illinois, which harked back to his involvement in the Civil War. Jenney died on 15 June 1907 in Los Angeles, California. His legacy remains in the North American skylines, particularly Chicago, along with the work of the many building designers he mentored and inspired.

TOWERING GENIUS

Another famed student from Paris' École Centrale des Arts et Manufactures was Gustave Eiffel (1832–1923). Eiffel designed the wrought-iron skeleton for the Statue of Liberty, a gift from France to the United States in 1886. He then leapt to fame for the tower he designed for the hundredth anniversary of the French Revolution in 1889. Despite an initial antipathy from locals, the Eiffel Tower became Paris' most famous symbol.

LEFT: Construction of the Eiffel Tower took two years, using 6,600 tonnes of wrought iron. The tower was only meant to stand for 20 years before being dismantled, but its popularity has kept it in place for more than a century.

GOTTLIEB
DAIMLER

'THE BEST OR NOTHING AT ALL.'

Translation of Daimler motto 'Das Beste oder nichts'

GREATEST ACHIEVEMENTS

INTERNAL COMBUSTION ENGINE 1883

REITWAGEN (FIRST MOTORCYCLE) 1885

FOUR-STROKE ENGINE 1889

FIRST MERCEDES 1900

ABOVE: Gottlieb Daimler.

Gottlieb Daimler, with his partner Wilhelm Maybach, helped design and build the world's first four-wheeled petrol-driven vehicle and established the automobile industry as we know it today.

Born in the village of Höllgasse in Schorndorf, Germany, on 17 March 1834, Gottlieb Daimler could easily have followed in his father's footsteps by becoming a successful baker. Instead, at school, he developed an interest in mechanics. Aged 14, he began four years of training as an apprentice gunsmith before progressing to the Stuttgart School for Advanced Training in the Industrial Arts, to study engineering. He was an earnest student, attending extra classes on Sundays. With help from his tutor, Daimler got involved in railway locomotive production, but it was clear to him that steam engines were on the way out.

Daimler moved to England to work in the Coventry factories of Sir Joseph Whitworth for a time, before relocating to France and Belgium. Finally, he returned to Germany in 1863, taking on the job of workshop inspector at the Bruderhaus Reutlingen engineering factory. It was here that he met the orphaned teenage engineer Wilhelm Maybach, who he would end up working alongside for many years to come.

In 1872 Daimler was taken on as technical manager at the Gasmotoren-Fabrik Deutz factory, the manufacturer of Nikolaus August Otto's hugely successful four-stroke internal combustion engines (see page 108). Maybach joined him there as chief designer. Daimler spent ten years at Deutz, never taking a holiday but he fell out with the company's management over engine designs. Daimler failed

BELOW: Daimler and Maybach's Reitwagen (1885) is considered by many to be the world's first motorcycle. It was based on a wooden bicycle with an internal combustion engine placed below the seat to drive the wheels.

ABOVE: Despite his fascination with designing petrol-driven vehicles, Gottlieb Daimler is said to have hated driving. In this photo from 1886, it is his son, Paul, taking the wheel in the first four-wheeled Daimler car.

to convince Nikolaus Otto of the potential of using lightweight four-stroke engines to power vehicles.

Both Daimler and Maybach left to set up their own business in 1882. Working from a converted greenhouse in Daimler's back garden in Cannstatt, the pair built a functioning petrol-powered engine of their own, with a carburettor that would mix the right proportion of petrol and air for combustion. Their noisy activity alerted neighbours who were convinced counterfeiters were at work, leading to a pointless police raid.

In 1885, Daimler and Maybach were ready to take their first vehicle out for a spin. Their 'Reitwagen' (riding car) was a wooden bicycle converted to carry an internal combustion engine. Daimler nicknamed the engine the 'Grandfather clock' for its resemblance to the pendulum-powered timepiece. On its test run, Maybach managed to drive the Reitwagen for 3 km (1.9 miles) and reach a speed of 12 km/h (7.5 mph).

In Mannheim, just 100 km (62 miles) away from Daimler and Maybach's workshop, another engineer, named Karl Benz (see opposite) was constructing an automobile of his own. Benz's 'Motorwagen' confirmed that a petrol-powered vehicle was possible. Daimler promptly ordered a four-seater horse-drawn carriage and set to work installing his engine and a steering wheel in it. The 'Grandfather clock' engine turned the carriage's rear wheels via a belt system and powered the world's first four-wheeled petrol-powered vehicle to a top speed of 16 km/h (10 mph).

Daimler and Maybach then proceeded to test their engine in other vehicles including a trolley car, an airship and a boat. The pair's success led to plentiful orders for vehicle engines, particularly for boats. They set up a factory outside Cannstatt and continued to develop road transport. In 1889, they launched the Stahlradwagen ('jet wheel cart'), a two-seater automobile closer to a tricycle than a car in design.

With demand increasing, Daimler and Maybach found backers and set up the Daimler-Motoren-Gesellschaft (DMG) in 1890 and, within two years, they sold their first automobile. Shortly after, Daimler collapsed with heart problems and had to take time off. When he returned, he faced boardroom battles at DMG. After failing to buy enough shares to take control, he sold his shares and patents, and quit. The ever-loyal Maybach followed him.

Daimler and Maybach continued their partnership, installing a new four-cylinder engine with Maybach's spray-nozzle carburettor in a vehicle they entered in the first organized automobile race from Paris to Rouen, France. The engineers must have gained some satisfaction when their entry won, beating all the automobiles entered by DMG. So it was that both Daimler and Maybach were made a huge offer to return to their former company. In 1895, DMG celebrated building their 1,000th engine. Daimler

engines were being licensed in Britain, France and the United States. The petrol-engine-powered automobile was a huge success and the world, for good or bad, was changed forever.

Daimler died from a heart condition on 6 March 1900. A month later, a new lighter model of Daimler car was completed, named after the daughter of a wealthy industrialist and automobile agent, the Mercedes.

At the start of the 20th century, competition was fierce between Benz & Co and their rivals, Daimler-Motoren-Gesellschaft (DMG) the maker of the Mercedes engine, now without their founder Gottlieb Daimler. In 1926, due to a faltering national economy, Germany's two largest automobile manufacturers agreed a merger, forming Mercedes-Benz.

KARL BENZ

Pioneer of automotive engineering, Karl Benz struggled through financial difficulties to invent the petrol-powered car but, with the help of his spirited wife, Bertha, he saw his designs give rise to the largest automobile company in the world.

Benz focused his attention on developing a petrol-fuelled two-stroke engine that he could patent. He invented many machinery parts now familiar to car drivers — the spark plug, throttle system, battery-powered ignition, gear shifters, carburettors, a water radiator and clutch. In 1885, he had managed to assemble these with a four-stroke petrol-powered engine in a three-wheeled vehicle, the Benz Patent Motorwagen. Benz's prototype was difficult to control and hit a wall during a public demonstration. For a time, Benz was banned from driving his invention in Mannheim, by the church, who saw it as the 'devil's carriage'.

Bertha Benz set out in the vehicle from Mannheim on 5 August 1888 for the 104-km- (65-mile-) journey to her mother's house in Pforzheim, taking her two sons with her. It was the first long-distance journey taken in a 'horseless carriage'. Before the dawn of petrol stations, the intrepid Bertha had to locate pharmacies to refuel the Motorwagen (petrol was sold as a household cleanser!) and deal with any mechanical problems. After several perilous downhill slopes, Bertha instructed a shoemaker to fit leather brake pads, and used a hat pin to clear a fuel blockage.

Incorporating Bertha's recommendations into the Motorwagen's design, such as adding an extra gear to deal with hills, and benefiting from Bertha's shrewd publicity stunt, Benz was ready to bring his vehicle to market. Going on sale in 1889, Benz's vehicle

was the world's first commercially available automobile. Six years later, Benz designed the first truck with an internal combustion engine.

By the end of the century Benz ran a thriving engine-manufacturing business, Benz & Co, with 430 employees. It was the largest automobile company in the world. Benz soon designed a cheaper automobile for mass production. The result was the Velo, a two-person automobile with a 2 kW engine and a top speed of 19 km/h (12 mph) .

BELOW: Benz designed the Velo (shown here with Benz driving and his wife Bertha as passenger) in 1893 as a less-expensive model that could be mass produced. It was the world's first large-scale production car.

THOMAS EDISON

GREATEST ACHIEVEMENTS

ELECTRIC VOTE RECORDER
Edison's first patent 1869

UNIVERSAL STOCK-TICKER
First big patent sale 1871

QUADROPLEX TELEGRAPH
Sends four messages down same line 1874

INVENTION FACTORY
Opens development lab, Menlo Park, NY 1876

CARBON TRANSMITTER
Improves telephone microphone 1877

FIRST VOICE RECORDING
Edison records 'Mary Had a Little Lamb' on phonograph 1877

INCANDESCENT LAMP
Improves carbon filament lamp 1879

ELECTRIC LIGHT DISPLAY
Pearl Street, New York City 1882

VITASCOPE
First projection of motion picture, New York City 1896

BATTERY POWER
Develops alkaline battery 1901

ABOVE: Thomas Edison.

'OUR GREATEST WEAKNESS LIES IN GIVING UP. THE MOST CERTAIN WAY TO SUCCEED IS ALWAYS TO TRY JUST ONE MORE TIME.'

Thomas Edison

A prolific inventor who illuminated New York City, Thomas Edison claimed 1,093 patents in his lifetime, with breakthroughs in sound, lighting and moving pictures.

The youngest of seven children, Thomas Alva Edison was born on 11 February 1847 in Milan, Ohio, USA. Known as 'Al', Edison was a hyperactive child and struggled to concentrate at school, so his mother chose to teach him from home. Edison showed an interest in chemistry and mechanics, as well as an early entrepreneurial flair. In his preteens, he worked on the Grand Trunk Railroad to Detroit, selling sweets and newspapers to passengers. Edison made use of the train's baggage car, installing a printing press to produce his own paper, the *Grand Trunk Herald*, and used the space as a chemistry lab. After Edison caused a fire the lab was swiftly shut down.

Aged 12, Edison lost most of his hearing, possibly as an after effect from a bout of scarlet fever, but it barely held him back. Indeed, in later life Edison claimed his poor hearing was a benefit, as he was less distracted and able to focus better on his work.

In 1862, Edison rescued a three-year-old boy from the train tracks before a boxcar could roll over him. In gratitude, his father, J. U. Mackenzie, introduced Edison to telegraphy, a communications technology used on the railroad. Months later, Edison was working as a telegraph operator in various US cities but, as telegraphy began using audible clicks rather

BELOW: A reconstruction of the laboratory at Thomas Edison's 'Invention Factory' at Menlo Park.

than printed Morse code, Edison's hearing problems became a handicap. In 1868, with his father out of work, and mother struggling with mental illness, Edison took employment in the Boston office of Western Union but continued to invent in his spare time.

A year later, Edison claimed his first patent, for an electric vote recorder. Unfortunately, politicians were reluctant to accept a device that sped up the process of voting. Edison determined not to waste time on unwanted designs from then on. Moving to New York City, he made his first major sale with a ticker-tape machine reporting on the stock market. The Gold and Stock Telegraph Company paid him for the rights. Now, Edison could afford to be a full-time inventor.

Edison opened his 'Invention Factory' in 1876. This large research centre in Menlo Park, New Jersey was an assembly line for innovation. Edison expected his employees to match his endurance by working long hours on experiments, and new patents were accumulated at a fast pace.

Within a year, Edison had developed a carbon transmitter for Alexander Graham Bell's telephone (see page 124), which improved the volume of a call. This led to his invention of the phonograph, a sound-recording device. The phonograph duplicated sounds by etching them with a needle on to a foil-coated cylinder, then played them back through a diaphragm. Edison tested

it by reciting the nursery rhyme 'Mary Had a Little Lamb' into the device. Despite its novelty, the Edison Speaking Phonograph was not a big success as it could only play back recordings a few times. It would be another decade before Edison made the most of its potential, for entertainment. Now, he was already looking at another medium – lighting.

Setting up a new factory in East Newark, in 1881, and moving his family close by, Edison set to deliver a useful electric light bulb. Using an existing bulb and spending 14 months on testing, Edison settled on a filament made of carbonized bamboo, as the best medium for a long-lasting electric-light source. An

initial display of 400 carbon-filament light bulbs in Manhattan's financial district led to over 10,000 orders within a year. Edison set up numerous companies to keep pace with the demand for his electric-lighting systems, and toured his illuminations around the world. He also helped provide the energy for electric lighting, building 12 power stations across the USA.

Unsurprisingly, Edison had his competitors. One lighting producer, George Westinghouse was insistent that his electrical system, using Alternating Current (AC) was better and could travel further, but Edison, who advocated Direct Current (DC), took him to task, claiming AC was more likely to cause lethal electric shocks. In the end, Edison backed the loser, with AC becoming the dominant current for general use.

Months after the death of his first wife, Mary, in 1884, Edison remarried, supposedly proposing to the 19-year-old Mina Miller using Morse code. He set up another, even bigger laboratory in New Jersey, large enough for 5,000 employees and equipped with a machine shop, phonograph and photographic departments, and a library.

After rival inventors Chichester Bell and Charles Sumner Tainter found success with an update of Edison's phonograph, using a wax cylinder and floating stylus, Edison spurned an offer of partnership and put his mind to improving his design further. After early attempts to market the phonograph as a business dictation device, in 1896, Edison launched the National Phonograph Co. and promoted his invention as a music player for the home. By 1912, the first fragile wax cylinders were replaced by sturdier materials and finally discs.

Two years after Edison launched the phonograph for the home, he announced, 'I am experimenting upon an instrument which does for the eye what the phonograph does for the ear.' This was the kinetoscope. One of Edison's colleagues, William Dickson, came up with the breakthrough – recording images on a celluloid strip, which could then be fed through a peephole viewer. This was the origin of motion pictures. The entertaining early films were of parades, dancers and boxing matches.

Surprisingly, Edison failed to see the potential for producing a motion-picture projector. It was Dickson, again, who developed the device. Several companies began competing for the market, even Edison, who caught up by delivering his 'Vitascope'. Later on, Edison tried, with minimal

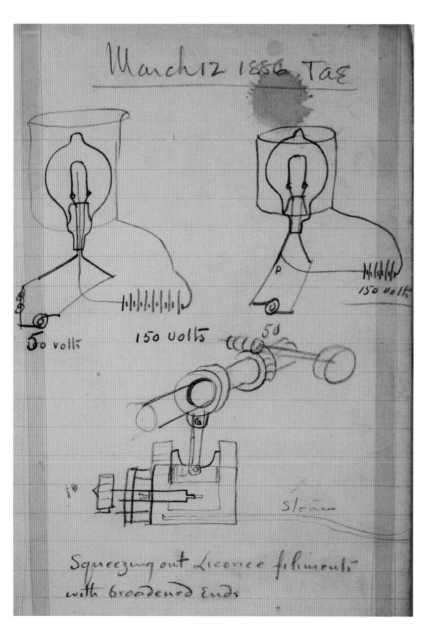

LEFT: Edison's designs for a lightbulb.

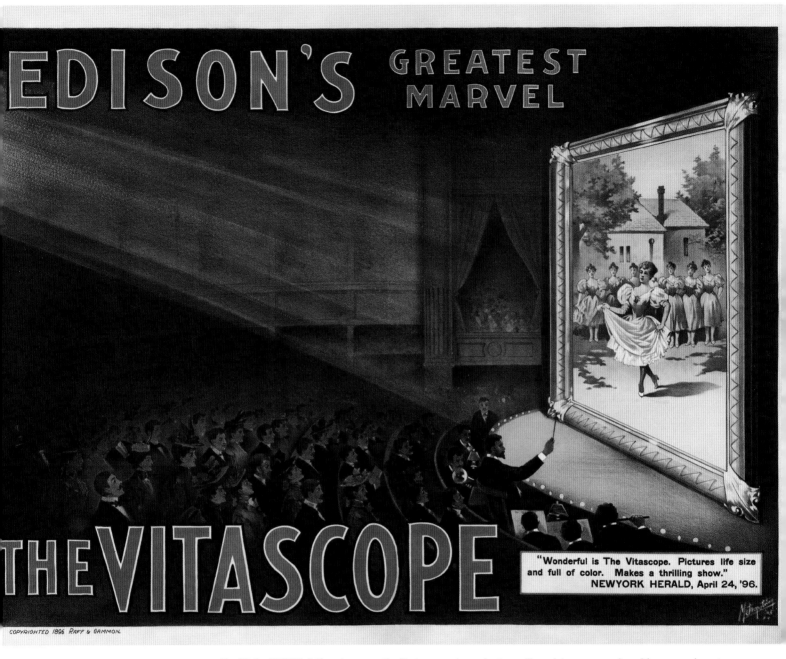

EDISON'S GREATEST MARVEL

THE VITASCOPE

"Wonderful is The Vitascope. Pictures life size and full of color. Makes a thrilling show."
NEWYORK HERALD, April 24, '96.

COPYRIGHTED 1896 RAFF & GAMMON.

ABOVE: On 23 April 1896, Edison became the first person to project a motion picture, or movie, with a screening at Koster & Bial's Music Hall in New York City.

success, to synchronize sound with his movie images.

Not every project Edison put his mind to was a triumph. In 1899, he set up the Edison Portland Cement Company, convinced cement was the ideal material for constructing low-cost homes, with concrete adaptable for use in furniture and even pianos. Neither worked out.

In 1911, Edison gathered all of his different companies under one umbrella, Thomas A. Edison, Inc. Four years later, World War I broke out and Edison stepped forward as head of the Naval Consulting Board to recommended technological defences, particularly submarine detection. He refused to develop weapons.

From the 1920s, now in his eighties, Edison's health deteriorated. On 14 October 1931, he fell into a coma due to complications with diabetes, and died four days later. His last breath is said to be kept in a test tube at the Henry Ford museum in Detroit. Street lights were dimmed to commemorate the life of the man who did so much to bring light to the world.

ALEXANDER GRAHAM BELL

'I THEN SHOUTED … THE FOLLOWING SENTENCE: "MR. WATSON, COME HERE. I WANT TO SEE YOU." TO MY DELIGHT, HE CAME AND DECLARED THAT HE HAD HEARD AND UNDERSTOOD WHAT I SAID.'

Notebook of Alexander Graham Bell, 1876

GREATEST ACHIEVEMENTS

ACOUSTIC TELEGRAPH (TELEPHONE) 1876

ESTABLISHMENT OF THE BELL TELEPHONE COMPANY 1877

ELECTRICAL BULLET PROBE 1881

GRAPHOPHONE
An improved version of the phonograph 1885

ABOVE: Alexander Graham Bell.

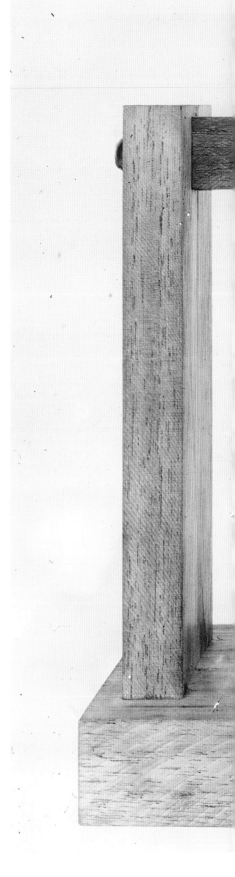

On 7 March 1876, the Scottish American inventor Alexander Graham Bell was granted a patent for an electrical communications device capable of 'transmitting vocal or other sounds'. His invention would change the world forever – it was the telephone. Securing the patent for the world's first practical telephone made Bell's fortune and guaranteed his place as one of the most famous engineers in history.

However, it had been a close race to the patent office. Elisha Gray, another US inventor, had independently developed a similar device. Just hours after Bell filed his patent on 14 February 1876, Gray submitted his design. A patent protects an invention from being copied and enables inventors to profit from their work. With so much money at stake, a bitter legal dispute over who was first and should own the rights to the telephone ensued. The case was eventually settled in Bell's favour, but it was just one of about 600 patent disputes that he would have to contest over the following years. Historians have argued about the justice of Bell's patent ever since, but his achievement in developing the first telephone is indisputable. The contributions of other innovators

RIGHT: A model of of Bell's first telephone of 1876.

THIS MODEL OF BELL'S FIRST TELEPHONE IS A DUPLICATE OF THE INSTRUMENT THROUGH WHICH SPEECH SOUNDS WERE FIRST TRANSMITTED ELECTRICALLY, 1875.

in the field, some of whose work Bell was familiar with, are better recognized today.

Bell was born in Edinburgh, Scotland on 3 March 1847. Inspired by his father, an elocution expert who had devised a system to categorize human speech sounds, the young Bell grew up similarly fascinated by sound and speech. By the age of 12, he was already something of an inventor, coming up with a mechanism to de-husk wheat. Inspired by a talking automaton, Bell also built a 'talking head' that could mimic simple speech-like sounds.

Bell would go on to become a teacher of the deaf, but he continued with his passion for building gadgets. Taking a lead from the German physicist Hermann von Helmholtz, Bell went on to experiment with ways to turn electricity into sound. Von Helmholtz had used electrical oscillations to transmit sound along a wire by making tuning forks vibrate. Bell initially misread Von Helmholtz's work, but his error led to a breakthrough. He realized that if it was possible to turn electricity into vibrations, the reverse would also be true. He also thought that if multiple pitches of sound could be transmitted along a wire, it might be possible to reproduce the entire spectrum of the human voice.

Following the deaths of his brothers, Bell followed his family to Canada in 1870 and then moved to Boston, USA, to teach in 1871. He continued to experiment with electricity and sound, turning his attention to improving the electric telegraph. By the late 19th century, the telegraph had made it possible to communicate at lightning speed. But there was a problem: only one message could be sent down a telegraph wire at a time. With increasing demand for the telegraph and the enormous expense of putting up more wires, the ideal solution was to find a way to send multiple messages down the same wire. This process, known as multiplexing, was the goal of many inventors and electrical engineers, including Bell and Gray. They set about trying to build a harmonic telegraph, which instead of transmitting a single set of pulses in Morse code, would carry several messages encoded at different pitches. Bell

RIGHT: Alexander Graham Bell making a call at the opening ceremony of the long-distance telephone line from New York to Chicago in 1892.

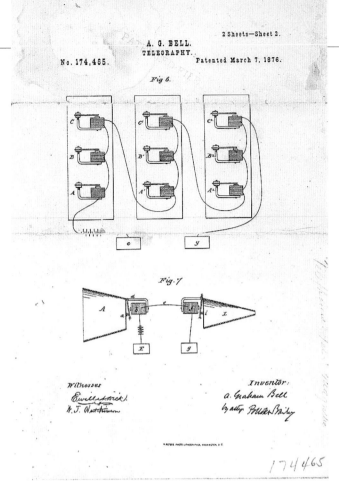

RIGHT: Bell's telephone patent was finally awarded following a bitterly contested legal battle with Elisha Gray.

and Gray were pursuing designs that would produce a range of tones at the receiving end. But Bell's experiments started to lead him towards the problem of transmitting human speech rather than simply improving the telegraph.

Bell had primarily been a teacher of the deaf, rather than an electrical engineer. Now, with his experiments at a critical stage, he needed expert help. Bell teamed up with Thomas A. Watson, a young technician employed at an electrical workshop in Boston. Watson had tackled all kinds of requests for electrical devices, and this provided him with the skillset to turn Bell's ideas into prototype devices. For his part, Bell brought together the various threads of investigation in the field with a unique perspective. Translating messages between different media had been the foundation of his work with the deaf, linking visual symbols and diagrams, to actions and spoken sounds. The same

ANTONIO MEUCCI

Some historians claim that Italian American inventor Antonio Meucci invented the first telephone in the 1850s. But whether his patent claims actually describe an equivalent electromechanical device remains uncertain.

LEFT: Antonio Meucci.

JOHANN PHILIPP REISS

In 1861, 15 years before Bell, the German physicist Johann Philipp Reiss had demonstrated his 'Telephon', a device capable of transmitting sound. Nobody realized its potential at the time.

OPPOSITE: Johann Philipp Reiss.

ELISHA GRAY

American inventor Elisha Gray was developing his own device to electrically transmit human speech along the same lines as Bell in the 1870s. It remains unclear just how much Bell and Gray knew of each other's work or patents.

OPPOSITE, RIGHT: Elisha Gray.

principles of translation underlay his work on the telephone. Bell was getting closer to translating the sound vibrations of speech in the air into undulating electrical currents in a wire. He had the vision, while Watson provided the technical know-how.

In the early 1870s, Watson constructed several prototype transmitters and receivers for Bell's harmonic telegraph using resonating metal strips instead of Von Helmholtz's tuning forks. Meanwhile, Bell's experiments for capturing speech sounds included using a dead person's ear to pick up vibrations and turn them into electrical signals! Bell was under increasing pressure from his financial backers to finish and patent the harmonic telegraph and leave his more ambitious experiments for later. On 2 June 1875, Watson and Bell were back working on the harmonic telegraph. While Watson was making adjustments to the metal resonators, Bell heard the spring twang, like an actual sound transmitted along the wire rather than a musical tone. The pieces of

the puzzle were coming together! Soon the metal strips of the transmitter had been replaced with a flexible membrane, wired up to convert vibrations into electrical signals. The harmonic telegraph was becoming a telephone. On 10 March 1876, after months of refinements, Bell's telephone successfully transmitted its first voice message. The story goes that Bell had spilled some acid and spontaneously called to Mr Watson for help. His assistant heard Bell's voice over the receiver, saying, 'Mr. Watson, come here. I want to see you.'

Bell's basic telephone would go on to be improved by many other engineers, including Thomas Edison, but Bell would be immortalized as its originator. The telephone would also overshadow Bell's many other engineering achievements, including designing a type of fax machine, an optical telephone called the photophone, a record-breaking hydrofoil boat, plus pioneering work on early aircraft and even an early air-conditioning unit.

VLADIMIR
SHUKHOV

GREATEST ACHIEVEMENTS

OIL PIPELINE (First in Russian Empire) 1878

CRACKING
Invents high-temperature oil-refining process 1891

ALL-RUSSIA EXHIBITION
Eight pavilions 1896

WATER TOWER
World's first hyperboloid water tower 1896

TRANS-CAUCASIAN PIPELINE
835 km (519 miles) 1906

ADZIOGOL LIGHTHOUSE
Kherson 1910

SHUKHOV TOWER
Hyperboloid radio tower, Moscow 1922

LENIN PRIZE
Awarded one of the Soviet Union's greatest honours 1929

ABOVE: Vladimir Shukhov.

'A THING THAT LOOKS BEAUTIFUL IS STRONG. THE HUMAN VISION IS USED TO NATURAL PROPORTIONS, AND ANYTHING ABLE TO SURVIVE IN NATURE IS ALWAYS STRONG AND USEFUL.'

Vladimir Shukhov

Little known outside his native Russia, Vladimir Shukhov was a master structural engineer, who used mathematical ideas to design graceful metal frameworks.

Vladimir Shukhov was born on 28 August 1853 in the small Russian town of Graivoron. His father, Gregory Petrovich Shukhov, had served as a Russian Army officer but was a local bank director by the time of Vladimir's birth. At his school, St Petersburg Gymnasium, Vladimir showed an aptitude in mathematics and graduated with distinction, set on a career in engineering.

On his father's recommendation, Vladimir enrolled at the Imperial Moscow Technical School where he studied physics and maths. He was a diligent student, spending hours in the reading room and workshops. Here, he came up with a new design for a steam injector used to help the combustion of liquid fuel. Earning a gold medal for his academic achievements, the young Shukhov turned down an opportunity to work as a researcher. He sought to put his engineering knowledge into practice instead.

In 1876, Shukhov joined a delegation headed to Philadelphia, USA, for the World Fair Centennial Exhibition. Here he met the Russian-American

BELOW: Shukhov designed eight pavilions for the All-Russian exhibition in Nizhny Novgorod, in 1896, incorporating the world's first membrane roofs.

ABOVE: Built during the Russian Civil War to broadcast propaganda in 1922, this hyperboloid radio tower designed by Shukhov was used until 2002. It was rescued from demolition in 2014.

entrepreneur Alexander Bari, who would have a huge influence on the young engineer's career. Bari was responsible for the construction of several buildings at the fair, helped source equipment for workshops and introduced the Russian delegation to US methods of railroad construction at the Pittsburgh metal factories.

Inspired by the overseas trip, Shukhov returned to Russia and got a job with the Warsaw–Vienna railway, helping design stations and depots. The work gave Shukhov little opportunity to be creative so he quit to join the Academy of Military Health. If not for the fortuitous return of Alexander Bari, Shukhov's engineering ambitions might have ended at this point.

Bari moved to Russia in 1877, expecting to benefit from huge industrial progress in the country. Given the role of head engineer for an oil business, Bari remembered the talented engineer he met in Philadelphia and invited Shukhov to run the company's office in Baku, Azerbaijan. Three years later Bari started his own construction firm and boiler fabrication plant. He signed up the young Shukhov as chief engineer and designer. The two of them would

work together for the next 35 years.

Shukhov designed the first oil pipeline in the Russian Empire. Opened in 1878, at 12 km (7 miles) in length joining Balkhany to Cherny Gorod, within five years it was part of a system of 94 km (58 miles) of pipeline Shukhov planned around Baku. In 1906 this figure was surpassed by a new Trans-Caucasian pipeline he designed that stretched an astounding 835 km (519 miles).

In 1891, Shukhov patented a new thermal cracking process for oil refineries, which heated oil to high temperatures to break down hydrocarbon molecules into simpler forms. This produced more useful fuels. Shukhov's designs for steam boilers, oil reservoirs and oil-tanker barges were brought into service across the whole of Russia.

Outside of the oil industry, Shukhov designed some of the most striking and inventive towers in the country, built with a latticework of metal beams. 1896 was a key year for the engineer. For the All-Russian exhibition in Nizhny Novgorod, which demonstrated the country's greatest technical and industrial achievements, Shukhov designed eight huge pavilions with a membrane roof and a steel grid shell. His 32 m (100 ft)-tall water tower was a highlight of the show and became the prototype for thousands more around the world.

Alexander Bari died in 1913 and his son moved to the United States rather than take over his business in Russia. World War I broke out a year later, followed by the Russian October Revolution, which saw many entrepreneurs flee the country. Shukhov chose to stay and help rebuild his homeland. He was commissioned by Lenin, the head of the new Russian Soviet government, to construct the Shabolovka Radio Tower in Moscow, for broadcasting the government's message to the people. Now known as the Shukhov Tower, it could have reached a height of 350 m (1,150 ft), much taller than Paris' Eiffel Tower, and a third its weight. Due to lack of resources, a more modest, yet still imposing tower was erected. Constructed with six hyperboloid structures winched up like an extending telescope, at 152 m (500 ft) high, the tower remained the tallest structure in Russia for many years.

In total, Shukhov oversaw the erection of around 200 towers and 500 bridges in Russia. He also designed pylons, railway stations, lighthouses and a rotating stage for a theatre in Moscow.

In his later years, Shukhov kept out of the limelight, only meeting close friends and colleagues at home. After an accidental fire left him with severe burns, he died on 2 February 1939 in Moscow. A deservedly famous name in Russia, Shukhov was honoured with the Lenin Prize in his lifetime, and, later, a university in Belgorod was named after him.

BELOW: The roof of Kievsky Railway Station in Moscow was designed by Shukhov. His lightweight, metal beam structures were years ahead of their time.

HERTHA AYRTON

'AN ERROR THAT ASCRIBES TO A MAN WHAT WAS ACTUALLY THE WORK OF A WOMAN HAS MORE LIVES THAN A CAT.'

Hertha Ayrton

GREATEST ACHIEVEMENTS

SPHYGMOGRAPH
Invented device to record pulse 1877–81

LINE-DIVIDER
Patented tool for equally dividing line 1884

INSTITUTION OF ELECTRICIAL ENGINEERS
First female member 1899

THE ELECTRIC ARC
Ayrton's research published as a book 1902

ROYAL SOCIETY
First woman to present a scientific paper 1904

AYRTON FLAPPER FAN
Anti-poison gas device used in World War I 1917–18

ABOVE: Hertha Ayrton.

With breakthroughs in the study of the electric arc and air currents, British engineer and inventor Hertha Ayrton broke through barriers in the male-dominated scientific community and inspired generations of women to study and take credit for their discoveries.

Hertha Ayrton was born Phoebe Sarah Marks in Portsmouth on 28 April 1854. Her father, Levi Marks, was a watchmaker who had fled from Poland to England to avoid anti-Jewish violence. He died when Sarah was just seven. Sarah helped her mother raise her other seven siblings until, aged nine, she moved in with her aunts in London. Her aunts ran a school and helped with Sarah's education. She showed an aptitude in science and maths, but also studied French and music in readiness for work as a governess. She took on this work from the age of 16 in order to send money to her mother.

Sarah's friends nicknamed her 'Hertha' after the heroine in a poem by Algernon Swinburne. Knowing Hertha was bright, they encouraged her to take the entrance exams for Cambridge, which had just started allowing girls to enrol. So it was, from 1877, that Hertha began studies at Girton College, the first university college for women. While a student, Hertha invented the sphygmograph, a device for recording the human pulse. Another of her early creations was the line-divider, a method for dividing lines into equal parts, a tool that would prove useful to architects and engineers.

Hertha graduated with a third-class certificate in mathematics in 1881. (Women were not awarded academic degrees by Cambridge University until 1948!) In 1884, she began attending evening classes at Finsbury Technical College given by the electrical-engineering expert Professor William Edward Ayrton. Within a year Hertha and her professor were married. Hertha Ayrton acted as stepmother for William's four-year-old daughter, Edith, and had her own daughter, Barbara, in 1886. In addition to the childcare, Ayrton assisted her husband with his research. While studying the electric arc, Ayrton made an important discovery. The electric arc was commonly used to produce lighting in the late 19th century, but it often hissed and flicked on and off. Ayrton realized that this was caused by oxygen coming into contact with the carbon rods used to create the arc.

As a result of her breakthrough, Ayrton was invited to present her research to the Institution of Electrical Engineers. She was the first woman to receive the honour and be made a member. At the time, it was difficult for women to be recognized in the sciences. Despite Marie Curie being the discoverer of radium, it was her husband who had received the credit.

RIGHT: Ayrton designed a special fan to waft poison gas away from troops during World War I. The fan could be folded and carried behind a soldier's main pack.

Ayrton was a close friend of Curie and supported her campaign to be given the credit she deserved.

There was no disputing the achievements of Hertha Ayrton, however. From 1883 until her death in 1923, Ayrton claimed 26 patents, including five for mathematical dividers and 13 for work on arc lamps and electrodes. In 1902, she was nominated to join the prestigious Royal Society but was turned down as married women were considered ineligible.

As her husband's health deteriorated, the family moved to Margate on the Kent coast, hoping the sea air might aid his recovery. While here, Ayrton started to observe the ripples caused by waves on the beach. This led to years of research into air and water motion, that continued beyond William Ayrton's death in 1908.

After war broke out in 1914, Ayrton realized her studies could help troops on the front line. New, deadly gases were being used as weapons, including mustard gas, which attacked its victims' lungs. Ayrton designed a special fan, made of canvas over hinged canes, to waft the gas away from the trenches, following her specific instructions. Initially her proposals were dismissed by the War Office until newspapers reported on her work and 104,000 fans were supplied to soldiers.

After the war, Ayrton continued to use her knowledge of air currents and vortices in an attempt to clear the noxious gases issuing from mines and sewers. In 1918, a cause very close to her heart succeeded in its aims when women over 30 were given the vote. Throughout her life Ayrton had joined marches and demonstrations, supporting the suffragette movement, and gained recognition for her scientific achievements. She led the way for future generations of women to study and find acceptance and success in science and engineering. Hertha Ayrton died of septicaemia on 26 August 1923.

NIKOLA TESLA

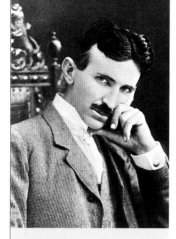

'ALL THE WORLD'S HIS POWER HOUSE.'

Time magazine celebrating Tesla's 75th birthday, 1931

GREATEST ACHIEVEMENTS

INDUCTION MOTOR
AC generator 1883

TESLA ELECTRIC COMPANY
Founded 1887

TESLA COIL
Resonant transformer circuit 1891

NIAGARA FALLS
Designs hydroelectric plant 1893

NEON LIGHT, X-RAYS
Takes first X-ray picture 1893–4

TELEAUTOMATON
Demonstrates remote-control boat 1898

ABOVE: Nikola Tesla.

An eccentric but visionary engineer, Nikola Tesla helped develop the transmission of electrical power and invented remote control, among a host of life-changing inventions, but he was never able to preserve any financial success.

Born on 10 July 1856 in the Austro-Hungarian Empire (now Croatia), the son of an Orthodox priest, Nikola Tesla was a bright and obsessive child who could memorize books and logarithmic tables. At the age of 19, Tesla travelled to Graz in Austria to study electrical engineering at the Polytechnic Institute. He worked hard, with little sleep, passing almost twice the required number of exams and earning the highest grades. Tesla was clearly highly intelligent but, in his third year, he squandered the opportunity to study, gambling away his tuition fees, before dropping out.

LEFT: The induction motor was one of Tesla's first and most successful inventions. It led to huge advances in the generation and transmission of alternating current.

RIGHT: Nikola Tesla made a huge contribution to the advancement of electrical power distribution and claimed around 300 patents over his lifetime. He is seen here sitting in front of his famous 'Tesla Coil'.

In 1881, Tesla moved to Budapest and took a job at the national telephone exchange. Here, he conceived of an induction motor that would generate alternating current (AC), but it would be another two years before he could afford to build a model. By then, Tesla was working in Paris, designing dynamos and motors for the Continental Edison Company, one of the US inventor Thomas Edison's (see page 118) many operations.

Tesla was convinced that Edison would see the potential in his new motor and AC. He sailed to the States in 1884, arriving with just four cents and a letter of recommendation in his pocket, to demonstrate his machine to Edison. The American entrepreneur was set on direct current (DC) as the dominant current for electricity and was not impressed by Tesla's induction motor. Yet, he did see potential in Tesla himself, and offered him work and a generous reward if he could improve the capability of his DC power supply. According to Tesla, he did just as requested, but the reward failed to materialize. Regardless of the circumstances, Tesla quit.

Edison may not have taken on Tesla's invention, but rivals were happy to invest in it. Tesla joined forces with two business-minded partners to form the Tesla Electric Company in 1887 and designed an impressive AC motor and power system. After much good publicity, the company received a handsome $60,000 in licence fees for his patent. George Westinghouse took on Tesla's designs and ended up in an expensive 'war of the currents' with Thomas Edison, who continued to endorse DC. Alternating current eventually succeeded as the electric current of choice. It could travel further and at higher voltages than DC.

With the proceeds from his licensing deal, Tesla set up his own laboratory in New York City in 1889. Following on from German physicist Heinrich Hertz's discoveries of electromagnetic radiation, Tesla began experimenting with a transformer that increased electrical voltage. Tesla's design featured two metal coils, with an air gap between them, wrapped around an iron core. This 'Tesla Coil' could generate high voltages and frequencies that could be used to power new forms of lighting – neon and fluorescent, and X-rays. Tesla saw its potential for transmitting electrical energy through the air and earth without the need for wires. He demonstrated the concept in public, using his invention to switch on lights across a stage.

Tesla was making a name for himself, and was asked to advise on a power-generating system at Niagara Falls in 1893. Using Tesla's designs, Westinghouse Electric won the contract to install a hydroelectric plant at the site. In another public demonstration, at Madison Square Garden in 1898, Tesla showed off his 'teleautomaton', a remote-control that steered a boat. The control used radio signals to guide the vessel. The inventor tried and failed to sell his idea to the US military. It would be another 20 years before it was taken seriously.

Around the turn of the century, Tesla spent much of his time trying to build a more-powerful wireless transmitter, in a race to beat European radio pioneer, Guglielmo Marconi (see page 160). Tesla filed a

RIGHT: In 1900, Tesla demonstrated a 'magnifying transmitter', producing artificial lightning in his Colorado Springs laboratory. Despite appearances, Tesla was not in the room at the time. A double-exposure was used to place him there.

... my illustrious friend Sir William Crookes
whom I always think and whose kind
letters I never answer!

Nikola Tesla

patent for his radio-wave technology in 1897 but, after a lab fire destroyed much of Tesla's research, Marconi leap-frogged his rival and transmitted the first radio signal across the English Channel in 1899. Tesla persevered, at great financial cost, to prove his wireless transmission designs were effective, building a 57 m (186 ft)-tall tower at Wardenclyffe, New York in 1904. His plans came to naught. Tesla faced mounting debts, and the project was abandoned in 1906 and the tower demolished in 1917.

Following this huge disappointment, Tesla struggled to find funding for his research and moved offices several times until he became effectively bankrupt. From 1919 to 1922, he worked with several companies, developing a power-generating bladeless turbine, Vertical Take-Off and Landing and radar technology. Tesla became increasingly eccentric, obsessively feeding pigeons, showing signs of obsessive-compulsive disorder, and announcing a series of outlandish ideas, such as a motor powered by cosmic rays, a thought recorder and a 'death ray'!

On 7 January 1943, a maid found Tesla's body in his rented room at the Hotel New Yorker. He had died of coronary thrombosis. Tesla left huge debts behind him but, more importantly, had conceived the basis of the electrical power system, advanced radio technology and delivered hundreds of patents.

LEFT: Tesla spent years fighting for funds to erect a wireless transmission tower at his laboratory at Wardenclyffe, Long Island, NY, but the project had to be abandoned in 1906.

RIGHT: Tesla's patent for a Vertical Take-Off and Landing (VTOL) aircraft from 1928.

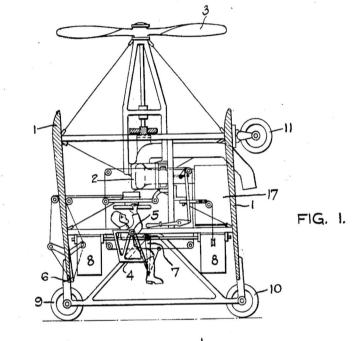

Jan. 3, 1928.   1,655,114

N. TESLA

APPARATUS FOR AERIAL TRANSPORTATION

Filed Oct. 4, 1927   2 Sheets—Sheet 1

FIG. 1.

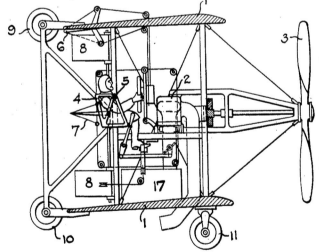

FIG. 2.

INVENTOR.

NIKOLA TESLA.

BY

ATTORNEY.

GRANVILLE
WOODS

GREATEST ACHIEVEMENTS

ESTABLISHED THE WOODS RAILWAY TELEGRAPH COMPANY 1884

TELEGRAPHONY 1885

SYNCHRONOUS MULTIPLEX RAILWAY TELEGRAPH 1887

BOILER FURNACE 1889

RE-ELECTRIC RAILWAY SUPPLY SYSTEM 1893

AUTOMATIC AIR BRAKE 1902

ABOVE: Granville Woods.

'MR WOODS IS ONE OF THE FOREMOST ELECTRICIANS OF THE COUNTRY AND HIS MANY INVENTIONS HAVE SHOWN AN AMOUNT OF SKILL AND DEXTERITY IN DEALING WITH THE MOST POTENT AND MYSTERIOUS OF ALL FORCES, WHICH PLACES HIM IN THE FRONT RANK OF INVENTORS ALSO.'

Cincinnati Commercial Gazette, 1889

The distinction between engineers and inventors is not always clear, and many engineers are remembered for landmark inventions rather than their engineering work. Often, engineers find solutions in existing technology and techniques, rather than inventing something new. That being said, the problem-solving skillset of engineers does make them uniquely placed to innovate. This was the case with Granville Woods, an African-American

OPPOSITE: Woods' 1884 patent for an improved boiler furnace of a steam locomotive.

BELOW: Telegraph key.

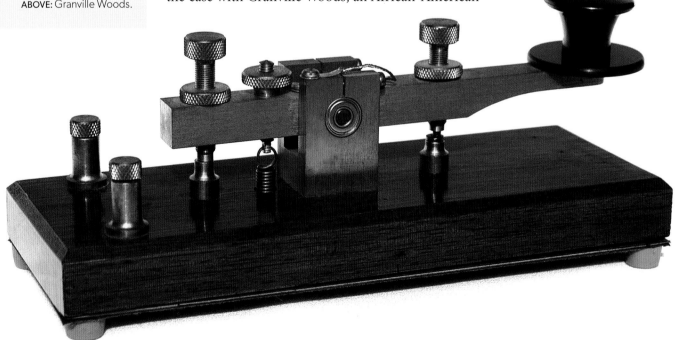

mechanical and electrical engineer who became a prolific inventor.

Working during the technological boom of the late 19th and early 20th centuries, Granville Woods took out more than 50 US patents for his inventions. Many were related to telecommunications and transportation, technologies that were rapidly developing as electrical power drove a second Industrial Revolution. Woods saw the potential in these emerging fields and was determined to pursue a career in engineering despite the obstacles in his way. As a black person in America just after the Civil War and the end of slavery, his opportunities for education and professional advancement were limited. Woods would have to struggle against racial prejudice throughout his career.

Granville Woods was born in Columbus, Ohio in 1856. His family were poor, so he had to leave school at the age of ten. He became an apprentice in a machine shop where he learned mechanical engineering and metal-working skills. Later, he found engineering-based jobs on the railroads, in steel mills and on a steamship. At some stage he also formally studied engineering, possibly at night school. By 1884, the largely self-taught Woods was ready to go into business. He set up the Woods Railway Telegraph Company with his brother, aiming to supply equipment and expertise to the telecommunications market. This had rapidly expanded since the 1830s, at first with the telegraph and then with the first telephone networks.

Woods' first patent was for an improved furnace for steam-train boilers, but it was his hard-earned expertise in electrical engineering that finally provided a breakthrough invention. He designed an improved telephone transmitter that combined the principles of telegraphy and telephones. He called his new system 'Telegraphony'. It allowed a single wire to carry either voice signals or telegraph-style Morse code messages. Alexander Graham Bell's telephone company promptly bought the rights to Woods' invention, if only to prevent this rival technology from competing with their growing telecoms empire. It was the first big financial success for the brothers' company. They used the money to pursue further research and development.

In 1887, Woods patented another electrical device – a Synchronous Multiplex Railway Telegraph. This allowed moving trains to communicate with stations and was an important invention that improved the safety of the rail network. Station managers now had a way to check the positions of their trains, reducing

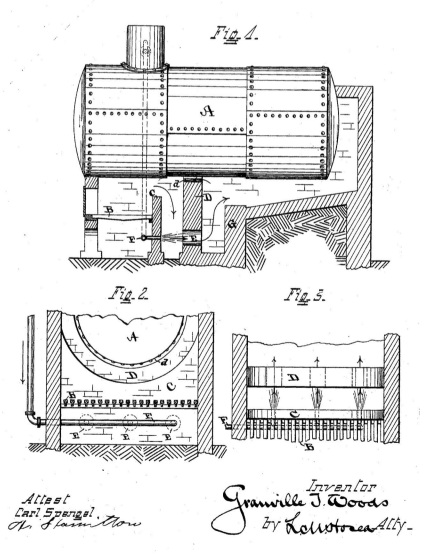

(No Model.)

G. T. WOODS.
STEAM BOILER FURNACE.

No. 299,894.

Patented June 3, 1884.

Fig. 1.

Fig. 2.

Fig. 3.

Attest
Carl Spengel.
A. Hamilton

Inventor
Granville T. Woods
by Lockhorea Atty.

G. T. WOODS.

INDUCTION TELEGRAPH SYSTEM.

No. 373,915.   Patented Nov. 29, 1887.

Fig. 1.

Fig. 2.   Fig. 3.

Fig. 4.

Fig. 5.

WITNESSES: INVENTOR

Granville T. Woods

By Kell Hosea

ATTORNEY

ABOVE: Patent for Synchronous Multiplex Railway Telegraphy.

now found himself up against the famous inventor Thomas Edison, who claimed that he had invented Synchronous Multiplex Railway Telegraphy first. Woods eventually won the case and Edison then offered him a job. Woods turned him down.

It was difficult for a black inventor to commercially exploit his own inventions, so Woods had to sell

the likelihood of crashes. Previous attempts to provide a communications link with moving trains had proved unreliable. They had relied on continuous contact with the telegraph lines alongside the track, but the irregular movements of trains often broke that connection and interrupted messages. Woods found an ingenious solution that avoided the need for continuous physical contact. He mounted an electromagnetic coil on the train that used a magnetic field to induce a current that carried messages along the telegraph wire.

Woods' invention brought public recognition of his talents, but it also led to conflict with rivals. Inventors often had to take costly legal action to protect their valuable patent claims in court. Woods his patents to make a living. But, with his financial resources depleted following costly court cases, Woods' later life was beset with financial worries and ill health. He died following a brain haemorrhage at the early age of 53 and was buried in an unmarked grave in New York. Despite a lifetime of engineering innovation, designing everything from egg incubators, to electric rails and automatic air brakes for trains, Woods died poor. Today, his remarkable engineering achievements have been rediscovered and finally recognized.

BELOW: A steam locomotive from the 1890s. Woods' Synchronous Multiplex Railway Telegraph significantly improved railway safety in the US.

RUDOLF DIESEL

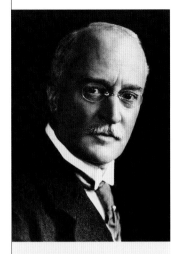

'I AM FIRMLY CONVINCED THAT THE AUTOMOBILE ENGINE WILL COME, AND THEN I CONSIDER MY LIFE'S WORK COMPLETE.'

Rudolf Diesel, 1913

GREATEST ACHIEVEMENTS

CLEAR ICE MANUFACTURE
First patent 1882

DIESEL ENGINE
Successful prototype tested 1897

SHIP ENGINE
Diesel engines first used at sea 1903

DIESEL TRAIN
First diesel locomotive 1913

DIESEL LORRY
First diesel-powered truck 1924

DIESEL CAR
Daimler-Benz produce diesel-powered cars 1936

ABOVE: Rudolf Diesel.

The diesel engine has been powering road, rail and water transport for almost a century, but its inventor disappeared in mysterious circumstances before he could witness its phenomenal success.

Rudolf Christian Karl Diesel was born on 18 March 1858 in Paris, France. His father, who had emigrated from Bavaria, was a bookbinder but struggled to support his family. As a child, Rudolf helped out in his father's workshop and made deliveries in a barrow. The young Diesel did well at school, earning a bronze medal for his studies at the age of 12. In 1870, war broke out between France and the Prussian Empire (Germany) and the Diesels had to leave their home, along with many Germans. The family relocated to London, England, but sent Rudolf to join his aunt in Augsburg and study German. While there, Rudolf decided he wanted a career in engineering and enrolled at the Royal Bavarian Polytechnic of Munich.

Diesel graduated from the Polytechnic in 1880 with the best grade ever awarded there. Having been enamoured by lectures he'd attended by the German scientist and engineer Carl von Linde, Diesel joined Linde in Paris to help him design a new refrigeration plant. Within a year, he had earned his first patent, for manufacturing clear ice, and been promoted as the factory's director.

In 1890, Diesel moved to Berlin, Germany, to run the research-and-development department of Linde's company. Using his comprehensive knowledge of

BELOW: German inventor and engineer Rudolf Diesel is remembered for the efficient engine and fuel that bears his name. Here, on the right, he is seen working on the engine that bore his name.

thermodynamics, Diesel experimented with constructing a fuel-efficient and thermally efficient steam engine running on ammonia vapour. But, during an early test, the engine exploded, nearly killing its inventor. Another test using high cylinder pressures resulted in another explosion. This time, Diesel ended up in hospital for months, with his eyesight severely damaged.

Despite the setbacks, Diesel continued his research, working on a test engine at the Augsburg Machine Works from 1893. It would take another four years for a satisfactory prototype to be built.

In 1897, Diesel unveiled his first working diesel engine, a 25-horsepower, four-stroke, vertical cylinder model. The design improved on the efficiency of earlier designs by injecting the fuel at the end of the piston's compression stroke. The fuel was then ignited by the high temperature delivered by the compression rather than needing a spark plug to set it alight. Diesel tested various fuels, including powdered coal, for his engine before settling on paraffin. The liquid fuel used in diesel engines would eventually be named after the engine's inventor.

The Augsburg factory began manufacture of the engine a year later, before it was truly ready. Early buyers were not entirely satisfied, and some models were returned with mechanical problems. Diesel attempted to rectify the problems reported by including a new atomizer and improving the air-compression method.

The first diesel engines were designed for stationary use. They began powering ships from 1903, but it would be another decade before versions were tested on locomotives, then for driving automobiles. Sadly, Rudolf Diesel would not live to see this happen.

The success of his efficient engine had made Diesel a fortune, but he was bothered by criticism of his role in its invention. On 29 September 1913, during

ABOVE: The original patent for the diesel engine.

an overnight crossing from Antwerp to London on the Post Office steamship *Dresden*, Rudolf Diesel disappeared. His cabin bed had not been slept in. His hat and watch had been left behind. Some suspect he committed suicide by jumping overboard, but conspiracy theories abound. Diesel's body was discovered in the North Sea ten days later. A sealed envelope left for his wife contained the modern equivalent of $1.2 million, though documents suggest the engineer was close to bankruptcy.

Had he lived, Diesel would have seen his dreams come to pass, with his engine powering countless automobiles. Variations of his initial engine design are still providing power for cars, lorries, trains and ships.

OPPOSITE: The first, stationary diesel engine was built by Rudolf Diesel at the Augsburg-Nuremberg engine works in 1893. It is now displayed in the factory's museum.

AUGUSTE & LOUIS
LUMIÈRE

GREATEST ACHIEVEMENTS

ETIQUETTES BLEUE
Louis invents dry photographic plate 1881

CINÉMATOGRAPHE
Movie projector 1894

1,400 MOVIES
Films recorded in ten-year period 1905

AUTOCHROME
Colour photography process 1907

ABOVE: Auguste and Louis Lumière.

'MY INVENTION CAN BE EXPLOITED FOR A CERTAIN TIME AS A SCIENTIFIC CURIOSITY, BUT APART FROM THAT IT HAS NO COMMERCIAL VALUE WHATSOEVER.'

Auguste Lumière underestimating the Cinématographe, 1913

Two French brothers brought lights, camera and action to the world with their invention of the motion-picture camera and projector.

Auguste and Louis Lumière were born in Besançon, France, two years apart (Auguste on 19 October 1862, Louis on 5 October 1864). Their father, Charles-Antoine Lumière, was a photographer who had turned his hand to producing photographic plates, but business was slow and, by 1882, seemed doomed to fail. Having studied optics and chemistry at technical school La Martinière, in Lyon, Auguste and Louis set about designing an automated process to produce their father's plates. Louis invented a new dry photographic plate, the 'etiquettes bleue' which used a gelatine emulsion on light-sensitive paper. This allowed photographers to take pictures, then develop their images later, rather than needing urgent access to a darkroom. It became a huge success. By 1894, the Lumières' business was booming, employing 300 staff at a factory in Monplaisir, producing 15 million plates a year.

After attending a demonstration of Thomas Edison and William Dickson's new kinetoscope (see page 122) in Paris, Antoine Lumière returned to Lyon to show his sons a strip of film used in the device.

BELOW: Auguste and Louis Lumière grew up surrounded by the photographic equipment used by their father and later invented successful methods for a new dry photographic plate and a way of capturing colour images.

The kinetoscope only allowed one person at a time to watch a moving-picture show through a small peephole. Antoine was convinced Auguste and Louis could do better and design a machine that could project moving pictures for a wider audience and one that was much lighter and cheaper too.

Auguste and Louis leapt at the challenge and, by 1894, they had come up with a movie projector they called the Cinématographe, which photographed, developed and projected moving pictures. It was the world's first complete, and portable, film camera. Their invention was much lighter than Edison's kinetoscope film camera, at 5 kg (11 lb). Operated using a hand-powered crank, the Cinématographe projected 35 mm-wide films at 16 frames per second, slower than Edison's 46 frames but smoother. The projector was also quieter and used less film.

A breakthrough in the Lumières' Cinématographe was the way the film passed through the machine. Inspired by the operation of a sewing machine, which paused briefly between each sewing action, Louis added sprocket holes to film strips. As they passed in front of the Cinématographe's lens, they paused while the camera's shutter opened and closed. In this way, the film would pass smoothly through the machine and each frame of the film would be exposed for just the right amount of time.

The Lumières' first public screening with the Cinématographe took place on 28 December 1895 at the Grand Café in Paris, organized by Antoine Lumière. The first 50-second film was simply footage of workers leaving the Lumières' factory, but it entranced its audience. A year later, the Lumières had opened theatres in London, Brussels and New York City, showing a range of documentary films and comedy. The portable film camera could be carried to far-off locations, such as Japan, North Africa and Central America, and was used to capture the crowning of the last tsar, Nicolas II in Russia. More than 1,400 moving pictures were made between 1895 and 1905. Many of these movies can still be watched today.

Having helped launch the cinema industry, the brothers turned their attention back to their first love, photography, and colour. Colour photography had been attempted before but not in any satisfactory way. The Lumière brothers used microscopic red, green and blue potato starch grains on photographic plates. The grains acted as a filter for light before it hit a photographic emulsion that could be developed into a transparent image to be viewed against a light. This Lumière Autochrome was

launched in 1907 and became another international success, with photographers able to capture full-colour images from their travels.

The Lumière Company continued to produce photographic material for many years, while the brothers turned to other areas to test their inventive skills. In the 1930s Louis began work in stereoscopy, trying to produce successful 3D imagery, while Auguste designed medical instruments and began research on cancer and tuberculosis. Louis passed away on 6 June 1948, Auguste on 10 April 1954. The photographic and cinema industry that they helped develop continues to dazzle and entertain audiences around the world.

RIGHT: The first public demonstration of the Lumières' Cinématographe in 1895 attracted only 30 people but, once word got out, thousands flocked to see the first 'movies'.

WILBUR & ORVILLE
WRIGHT

'THE DESIRE TO FLY IS AN IDEA HANDED DOWN TO US BY OUR ANCESTORS WHO, IN THEIR GRUELLING TRAVELS ACROSS TRACKLESS LANDS IN PREHISTORIC TIMES, LOOKED ENVIOUSLY ON THE BIRDS SOARING FREELY THROUGH SPACE, AT FULL SPEED, ABOVE ALL OBSTACLES, ON THE INFINITE HIGHWAY OF THE AIR.'

Wilbur Wright, speaking at an Aero-Club de France banquet, 5 November 1908

As the 20th century began, the dream of flight in a controllable heavier-than-air vehicle seemed finally within reach. Having watched aeronauts in balloons and airships lead the way for decades, aviation engineers were getting closer to building a stable powered aeroplane. The race was on to make the final breakthrough on both sides of the Atlantic. American and European engineers kept a nervous eye on their competitors as they refined and tested their designs. Among these early aviation engineers were

OPPOSITE: Rival aviation pioneer Samuel Langley's piloted version of the Aerodrome fails its test flight .

BELOW: The Wright brothers' 1900 Glider.

GREATEST ACHIEVEMENTS

DISCOVERY OF WIND WARPING 1899

WRIGHT KITE 1899

FIRST WRIGHT GLIDER 1900

FIRST FLIGHT TRIALS Kitty Hawk, North Carolina 1900

WIND TUNNEL 1901

WRIGHT FLYER 1903

FIRST POWERED FLIGHT 1903

WRIGHT FLYER 1905 The first practical aeroplane.

ABOVE: Wilbur (top) and Orville Wright.

two American brothers, Wilbur and Orville Wright. They would go down in history for achieving the first heavier-than-air aeroplane flight on 17 December 1903.

Wilbur Wright was born in 1867, four years before Orville. Their father was a bishop and the editor of a church newspaper in Dayton, Ohio. The Wright family was close-knit and the boys grew up independently minded. Wilbur was outstanding at school and was set to go to college, but a hockey accident knocked out his front teeth. While he recovered at home, Wilbur read avidly. The subject that most captured his imagination was aviation. As boys, the brothers had been fascinated by a helicopter toy that their father had brought home. Now Wilbur started to properly study the problem of flight.

Meanwhile, Orville had left school early and started a printing business. Wilbur eventually joined him and together they learned valuable engineering skills while manufacturing printing presses. With the arrival of the safety bicycle in America, Orville enthusiastically took up cycling. This led to the brothers' next engineering venture. In 1892 they set up The Wright Cycle Company, to rent, sell and repair bicycles. They went on to manufacture their own bicycles, incorporating several inventions of their own. The brothers' bicycle business was very successful and it provided them with the financial independence to pursue their growing interest in aviation.

The brothers had read newspaper accounts of various flying machines and their intrepid inventors. The tragic death of glider pioneer Otto Lilienthal highlighted the instability of current aircraft designs once they were in the air. Being familiar with the forces at work in balancing on a bicycle, Wilbur and Orville grasped that a glider similarly needed continuous adjustments by the pilot to keep it stable and steer it. From 1899, the brothers set about a systematic study of aircraft design. They made a spinning machine from an adapted bicycle to test the aerodynamics of various wing shapes. Model gliders were then built and tested before they moved on to full-size prototypes.

From 1900, the Wright brothers spent a season each year carrying out flight trials on the coastal sands near Kitty Hawk, North Carolina. When their initial glider designs failed to perform they went back to the drawing board, built a wind tunnel and tried out more ideas. By the September of 1902, the Wright brothers had improved the lift of their glider. They had also worked out a way to control its roll by using wires to

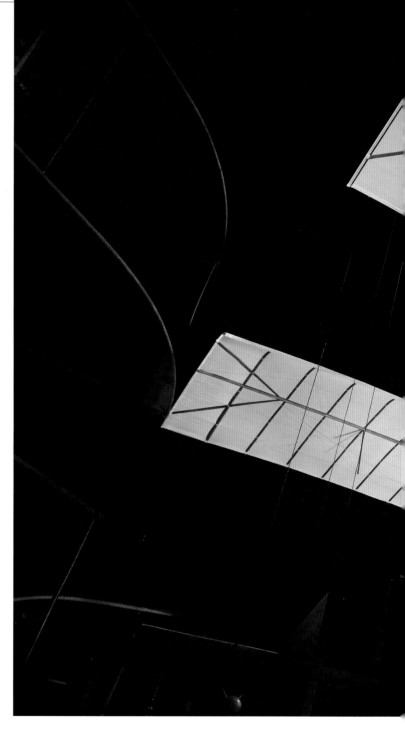

bend the edges of the wings. They called this 'wing-warping' and it performed the same job as ailerons on aircraft wings today. Together with a front elevator to control pitch and a rear rudder to control yaw, their glider now had three adjustable flight surfaces that provided control over the three axes of its movement. They took turns to make controlled flights of up to 190 m (623 ft) with their *Number 3 Glider*. Now all the brothers needed was a lightweight engine.

The Wrights opted for a gasoline-powered internal combustion engine. They couldn't source a suitable car engine, so their assistant Charlie Taylor built a bespoke lightweight engine. The race was on, because another American aviation engineer, Samuel Pierpoint Langley, looked ready to pip them to the post. He had built a small steam-powered model aircraft in 1896 called *Aerodrome Number 5*, which he claimed was the first heavier-than-air machine to fly. Langley's flying models had convinced the US Army to finance a full-scale version capable of carrying a pilot. As the Wright brothers headed to Kitty Hawk to test their first powered aircraft, the *Wright Flyer*, they knew Langley was poised to launch his *Great Aerodrome*.

They needn't have worried. On 7 October 1903, Langley's *Great Aerodrome* dropped straight into the Potomac River following its launch. A second attempt on 8 December was also a failure as the aircraft broke up during launch. The field was clear. On 14 December 1903 the brothers assembled their *Flyer*. Wilbur launched from a tall sand dune, but stalled and landed with a bump three seconds after take-off. Following repairs, the brothers tried again on 17 December, launching from a wooden rail running along the flat sands. They took turns to fly. Orville made a 12-second flight on the first attempt. But the fourth and final flight of the day, made by Wilbur, was the one that went into the history books. Covering a distance of 260 m (653 ft) and lasting 59 seconds, it was the world's first controlled and sustained flight by a heavier-than-air aircraft.

The brothers released a low-key statement about their flight to the press. They were wary of giving away the secrets of their success before securing patents and consolidating their claims with public demonstration flights. While they worked on perfecting the *Flyer*, the rest of the world was almost catching up. On the other side of the Atlantic, European competitors were sceptical of the Wright brothers' claim. In France, a rival 'first flight' was claimed by Alberto Santos-Dumont, who had made a 220-m (722 ft) flight on 12 November 1906 in a kite-framed aircraft called the *14-bis*. However, Wilbur Wright quashed all doubts when he arrived in France and made an accomplished demonstration flight close to Le Mans on 8 August 1908. The manoeuvrability of the more advanced *Type A Flyer* made it clear that the Wright brothers' claims were fully justified.

Following their achievement at Kitty Hawk, the Wright brothers were acclaimed around the world. They went on to build improved versions of the

LEFT: The first powered flight of the Wright brothers' aircraft at Kitty Hawk, North Carolina in 1903.

BELOW: President Taft meets the Wright brothers at the White House in 1909.

Flyer in 1904 and 1905 and tried to interest the US Army in buying it. They eventually signed a contract with the Wright Aircraft Company in February 1908 for an aeroplane capable of carrying a pilot and a passenger. This was the *Type A Flyer* that Wilbur then demonstrated in France. Sadly, the brothers wasted some of their head start in aircraft development, diverting too much energy to legal battles in a bid to protect their patents. Their production aircraft soon had competition from rival manufacturers who had caught up or had even overcome some of the shortcomings of the *Flyer*.

The protracted legal wrangling over patents took its toll on the brothers. Worn out by court cases, Wilbur died of typhoid fever on 30 May 1912. Soon after, in 1915, Orville sold his interests in the brothers' aircraft company and returned to research and development as a consultant aeronautical engineer. Orville continued to defend the brothers' claim to have built the first proper aeroplane against the Smithsonian Institution that championed Langley. The Smithsonian finally relented in return for the right to display the *Flyer*. Orville died in 1948, but not before he had seen the incredible dream of people flying in aeroplanes become an everyday reality.

GUGLIELMO
MARCONI

GREATEST ACHIEVEMENTS

WIRELESS TELEGRAPH
Patent for radio-wave system 1896

FIRST INTERNATIONAL WIRELESS COMMUNICATION
From France to England 1899

TRANSATLANTIC MESSAGE
From England to Canada 1901

NOBEL PRIZE
Awarded shared prize for physics (with Karl Ferdinand Braun) 1909

TITANIC
Marconi radio used to send S.O.S. 1912

ABOVE: Guglielmo Marconi.

'EVERY DAY SEES HUMANITY MORE VICTORIOUS IN THE STRUGGLE WITH SPACE AND TIME.'

Guglielmo Marconi

Italian inventor Guglielmo Marconi's work with the wireless telegraph led to the development of long-distance radio transmission and a dramatic rescue following one of history's most tragic events.

Guglielmo Marconi was born in Bologna, Italy, to an Italian aristocratic father and Irish mother on 25 April 1874. His parents were wealthy and provided him with a private education, tutoring him in mathematics, chemistry and physics. Though not enrolled, Marconi received permission to attend lectures at the University of Bologna. Here, Marconi became familiar with the German physicist Heinrich Hertz's work on electromagnetic radiation, or radio waves. In 1894, Marconi began his own experiments on the subject from his attic laboratory, attempting to use Hertz's discoveries to develop a communication system. The electric telegraph had been used for years to send Morse code messages over long distances. Marconi thought it possible to use radio waves to do the same, without wires.

By the end of the year, Marconi had assembled a radio transmitter and receiver to trigger a remote bell at home. By summer 1895, Marconi had moved outside and set up a tall antenna to broadcast radio signals over a distance of 2.5 km (1.5 miles). When Marconi approached the Italian Ministry of Post and Telegraphs with his wireless telegraph machine, they showed no interest. So, with letters of introduction, Marconi headed to England, where Sir William Preece, the chief engineer of the Post Office, encouraged him to patent his radio-wave communications system in 1896.

BELOW: Marconi saw the potential of electromagnetic radiation for sending and receiving messages, and led the development of international radio communication.

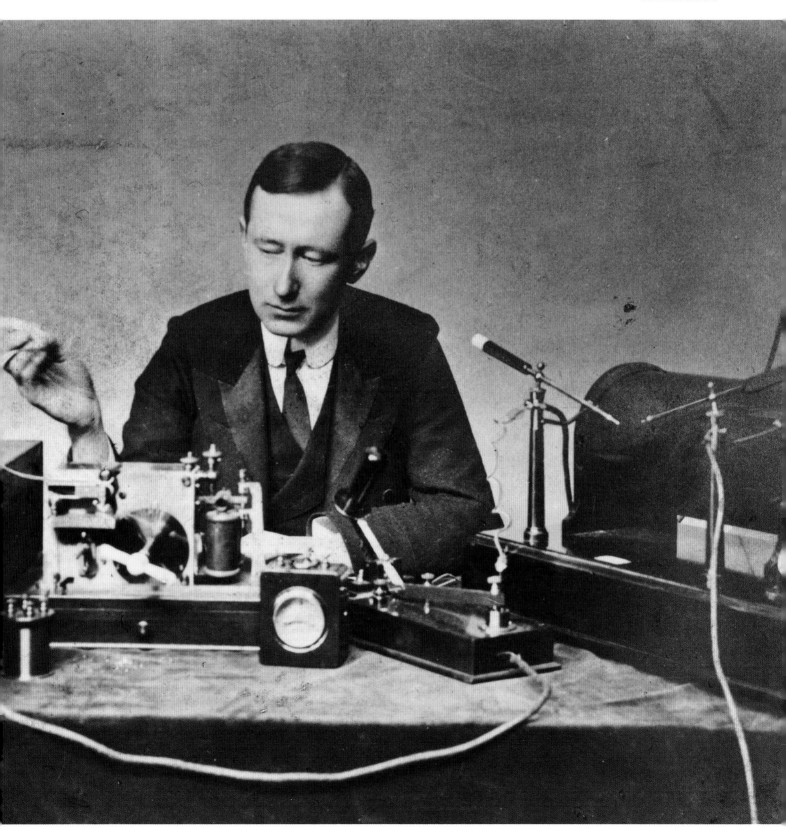

ABOVE: The world's first transatlantic radio message was transmitted in 1901 from this site in Poldhu, Cornwall, England, using four 65 m (213 ft)-tall masts.

Marconi began demonstrating the potential of his radio system to the British government. In 1897, he sent a Morse code message over 6 km (4 miles) across Salisbury Plain. On 27 March 1899 he sent a message across the English Channel from Wimeraux, France to South Foreland Lighthouse, Dover, England. Travelling 50 km (30 miles), this was the first international wireless communication. Later that year, to great acclaim, Marconi supplied equipment for two ships to report live to newspapers on the progress of the America's Cup yacht race.

It was assumed that 'straight' radio waves would not be able to travel around the curve of the Earth but Marconi tried it anyway. On 12 December 1901, Marconi and his assistant, George Kemp, set themselves up on a hill in St John's, Newfoundland

(now Canada) and received a signal all the way from Cornwall, England, 3,500 km (2,200 miles) away. The message was simply the Morse code for 'S', but it proved that radio waves could cross the Atlantic. Unaware that the waves were reflected off the Earth's upper atmosphere, Marconi was delighted when his attempt at long-distance wireless communication succeeded. Later tests delivered signals between Ireland and Argentina, and England and Australia.

The usefulness of wireless communication was made clear on 15 April 1912 when the RMS *Titanic* hit an iceberg and began to sink. Marconi's equipment was used to send out an S.O.S. signal, which was received by the liner *Carpathia*, three-and-a-half hours away. The *Carpathia* managed to rescue 705 passengers from *Titanic*'s 20 lifeboats.

In the 1920s, Marconi experimented using shorter wavelengths for his radio transmissions. The benefits of this were an increase in signal strength, more directed broadcasts and less opportunity for messages to be intercepted. In 1924 Marconi's company was rewarded with a contract to deliver short-wave communications between England and British Commonwealth countries. Marconi's company was also involved in early television transmissions in England.

Marconi married twice, to the daughters of aristocrats, enjoying a very comfortable and wealthy lifestyle. Returning to Italy in 1935, Marconi began actively supporting the fascist party of Mussolini. He died in Rome on 20 July 1937 after a series of heart attacks and received a state funeral.

Marconi was not the first to work with radio waves, and much of the technology he used was developed using other inventors' work. In a 1904 battle over the patent for inventing radio Marconi beat Nikola Tesla (see page 136) and went on to profit from supplying international radio communications. (The courts overturned the ruling in favour of Tesla in 1943, the year of Tesla's death.) By exploiting and developing others' work, Marconi enabled radio communications to become widespread and international. On his death, transmitters around the world were switched off to provide two minutes of silence in his memory.

BELOW: This large broadcast transmitter was the first to be operated in Britain, from the Marconi Works in Chelmsford, Essex, between 1919 and 1920.

LILLIAN MOLLER
GILBRETH

'THE MENTAL HEALTH OF THE WORKER NOT ONLY CONTROLS...HIS PHYSICAL HEALTH, BUT ALSO HIS DESIRE TO WORK.'

The Psychology of Management, 1914

GREATEST ACHIEVEMENTS

THE PSYCHOLOGY OF MANAGEMENT
Lillian's dissertation published 1914

HOUSEHOLD MANAGEMENT
Improvements in home economics 1929

PURDUE UNIVERSITY
First female engineering professor 1935

NATIONAL ACADEMY OF ENGINEERING
First woman elected 1965

HOOVER AWARD
From the American Society of Civil Engineers 1966

ABOVE: Lillian Gilbreth.

America's first lady of engineering, Lillian Moller Gilbreth was one of the first female engineers to be awarded a Ph.D. She used her knowledge of psychology to bring new levels of efficiency into the workplace, while considering the health of employees.

Lillian Moller was born on 24 May 1878 in Oakland, California, USA. She was the oldest of nine children in a well-off family, but rather shy and she had to be schooled at home for her first few years. Once she began at elementary school, Lillian struggled to make friends but showed a flair for music and verse. At the time, girls were not expected to continue on to college and careers. Lillian's parents expected her to settle down with a rich husband. Lillian had other plans and enrolled on a teaching course at the University of California, Berkeley. Here, she excelled in her studies on English literature and psychology and was given the honour of providing the commencement speech at the graduation ceremony in 1900.

Lillian met Frank Bunker Gilbreth, the wealthy owner of a Boston-based construction company, in 1903. The pair were married a year later. At the

suggestion of Frank, Lillian focused her studies on psychology so she might help manage his business. From then on, the couple worked together on finding the best, most-efficient working methods to improve productivity in the factory. Their insights would change working practices across the world.

Lillian and Frank went on to have a huge family, with 12 children. The story of their family life and the parents' attempts at efficiency in home life was later told by their children in two books, *Cheaper by the Dozen* and *Belles on their Toes*, both of which were turned into movies.

Lillian Moller Gilbreth was a pioneer in the world of industrial management. She considered the well-being of workers as well as the best use of time in her theories. She also improved and standardized factory tools and machinery, to make them easier to use. In 1913, the Gilbreths began teaching their ideas, opening the Summer School of Scientific Management for four years.

Lillian's dissertation and her defining work, *The Psychology of Management* was published in 1914. Earlier titles written by the Gilbreths on their management ideas failed to credit Lillian at the publisher's request. They thought having a woman's name attached would erode the authority of the books!

In the summer of 1924, Frank Gilbreth died unexpectedly of a heart attack. He was 55. Lillian never remarried and never gave up on their ideas. Chauvinistic attitudes in the engineering business made it difficult for Lillian to remain acting as a consultant for industry but she continued to lecture and found a new area in need of time-management improvements – the household. By simplifying domestic tasks and reducing the time they take, Gilbreth aimed to liberate women and allow them to take jobs in the wider world. Her work in this area, demonstrated at a Women's Exposition in 1929, led to better kitchen layouts, including time-saving technology. Simple improvements Gilbreth suggested included the pedal-operated bin and

ABOVE: Turning away from the factory, Gilbreth also improved the layout of kitchens, to save on time spent at home, even though she herself could barely cook.

shelves in refrigerator doors. Among Gilbreth's patented inventions were the electric tin opener and the washing machine wastewater pipe.

During the Great Depression of the 1930s Gilbreth worked as an advisor to President Hoover to help the unemployed find new work. She created a successful 'Share the Work' scheme. Gilbreth continued as a government consultant throughout World War II, helping convert factories for military needs.

Gilbreth died of a stroke at the age of 93, on 2 January 1972 in Phoenix, Arizona. The list of her awards is extensive, with 23 honorary degrees and, notably, the Hoover Medal for distinguished public service in 1966. She has been described as 'The World's Greatest Woman Engineer' for her impact on industrial practices, that benefit not just productivity but also the health and welfare of the workforce.

LEFT: Gilbreth not only recommended better use of time within factories but also improved upon tool design, to make them easier to use.

ROBERT H. GODDARD

'IT IS DIFFICULT TO SAY WHAT IS IMPOSSIBLE, FOR THE DREAM OF YESTERDAY IS THE HOPE OF TODAY AND THE REALITY OF TOMORROW.'

Robert H. Goddard

GREATEST ACHIEVEMENTS

MULTI-STAGE ROCKET
Patent approved 1914

VACUUM TEST
Lab test proves propulsion possible in vacuum 1915

BAZOOKA PROTOTYPE
Tube-based rocket launcher 1917

FIRST ROCKET LAUNCH
Powered by liquid fuel 1926

PAYLOAD FLIGHT
Rocket with barometer, thermometer, camera 1929

GYROSCOPIC CONTROL
Guided rocket flight 1932

HIGHEST ALTITUDE
Rocket reaches 2.7 km (1.7 miles) 1937

ABOVE: Robert H. Goddard.

Despite press ridicule, Robert H. Goddard pursued and succeeded in his dreams of launching a rocket with liquid fuel, and is now regarded as the foremost pioneer of modern rocket science.

Robert Hutchings Goddard was born on 5 October 1882 in Worcester, Massachusetts, USA, the only surviving child of a travelling salesman. Goddard was fascinated with science from a young age, and both his parents encouraged him by buying him a telescope and a microscope. A frail child, the young Goddard devoured science magazines and the science-fiction writings of H.G. Wells. By his teens, he was already experimenting using kites and gas-filled metal balloons.

OPPOSITE: American engineer Robert H. Goddard survived ill health and press mockery to prove that sending rockets into space was a possibility.

BELOW: Robert Goddard's rocket launcher from 1918 closely resembled what would become the bazooka.

In 1904 Goddard started at Worcester Polytechnic Institute where he impressed his physics teacher enough that he was invited to join him as a lab assistant and tutor. Among his imaginative ideas was a future where humans travelled in cars suspended in vacuum tubes and were shunted about by the power of electromagnetism.

Goddard's studies continued at Princeton University, where he earned a Ph.D. in physics. While there, Goddard presented his ideas on aircraft stabilization. His theories matched contemporary developments in gyroscopes (spinning devices which measure orientation and help pilots maintain a course). He expressed his ideas about rockets in 1909. After tests with solid fuel, he concluded that liquid fuel would be the best propellant as it was more efficient. Goddard proposed the use of liquid hydrogen, with liquid oxygen as the oxidizer to help combustion.

After a recovery from ill health, Goddard became a part-time instructor at Clark University. This allowed him time to pursue his experiments in rocketry. Despite assumptions that rockets would not work in the vacuum of space, where there is nothing to push against, Goddard proved in the university laboratory that propulsion would work without air. In 1914, Goddard registered two major patents, one for a multi-stage rocket design and one for a rocket

using liquid fuel. Both ideas laid the groundwork for decades of rocket design.

In 1917, Goddard received backing from the Smithsonian Institution for his rocket research, based on his report, *A Method of Reaching Extreme Altitudes*. With World War I underway, Goddard also provided weapons ideas based on his experiments. One was a plan for a cylinder-based rocket launcher, similar to what became the bazooka. The war ended before his design was put into production.

Goddard was years ahead of his time. In a letter to the Smithsonian in 1920, he raised ideas such as photographing planets using fly-by cameras on rockets, sending messages into space on etched metal disks and powering spacecraft using solar energy. The press found Goddard's ideas fanciful, though, and mocked them with headlines such as 'Believes Rocket Can Reach Moon'. From then on, Goddard chose to keep his progress from the public eye. Fellow scientists and engineers read his theories with much interest, however, including several in Germany, as it prepared for military conflict.

In November 1923 Goddard successfully tested a liquid-fuelled engine. After much refinement, he was ready to attempt the first liquid-fuelled rocket launch. On 16 March 1926, from his Aunt Effie's snow-covered farm in Auburn, Massachusetts, Goddard fired his

OPPOSITE: On 16 March 1926, Goddard fired his liquid-fuelled rocket 'Nell' 12.5 m (41 ft) into the air. This 2.5-second-long flight launched the age of rockets.

RIGHT: Moveable air and exhaust vanes helped keep Goddard's rocket on its course.

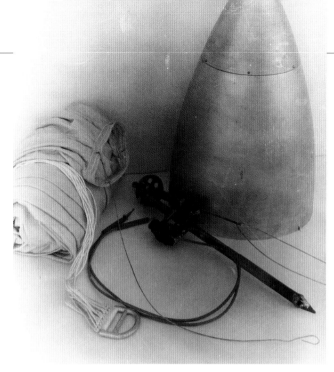

small rocket 'Nell', fuelled by petrol and liquid oxygen, 12.5 m (41 ft) into the sky. This humble beginning launched the rocket age.

Following his breakthrough, and thanks to the intervention of the aviator Charles Lindbergh, Goddard received significant funding from the Guggenheim family. Now, Goddard could afford to employ staff and lead test flights from a base in Roswell, New Mexico. In 1937, he fired a liquid-fuelled rocket that reached an altitude of over 2.7 km (1.7 miles). His continuing refinements of rocket design led to the introduction of fuel pumps, self-cooling motors and the use of gyroscopic guidance systems.

Goddard's offer of help during World War II led to him having to give up his Roswell base to work on jet-assisted take-off (JATO) engines for the Navy, from Maryland. Here, on the US East Coast, his health deteriorated. Goddard died of throat cancer on 10 August 1945. Despite some mockery and a lack of government backing, Goddard's work inspired and influenced those that followed him in the development of missiles and the US space programme. As a tribute, an asteroid and a crater on the moon are named after him.

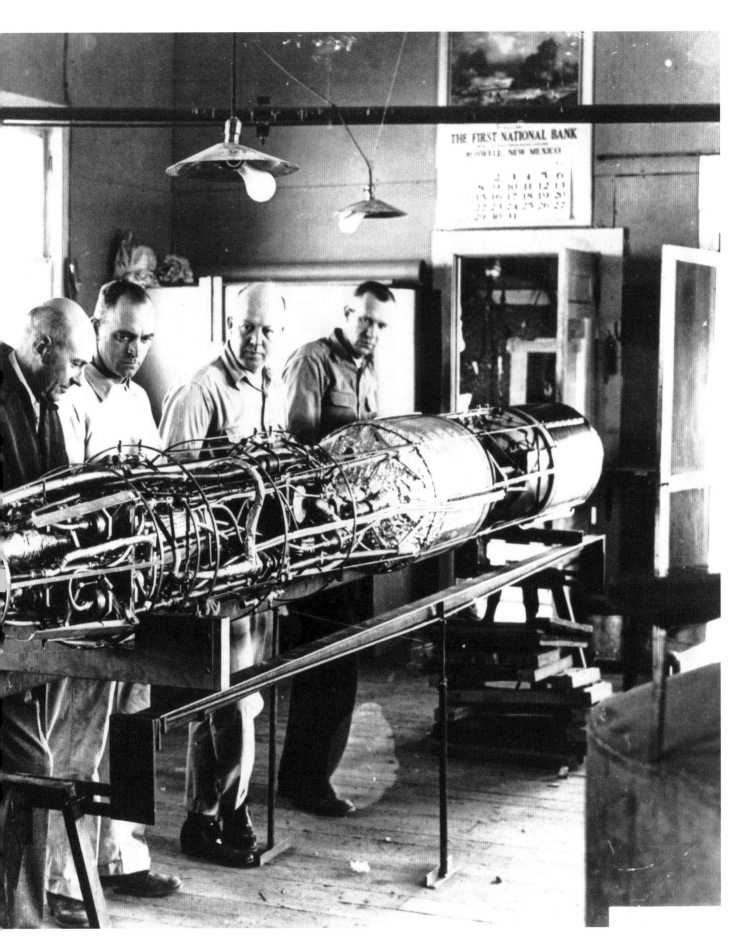

NORA STANTON BLATCH
BARNEY

ABOVE: Nora Stanton Blatch Barney.

'A TALENTED ENGINEER, ARCHITECT AND MATHEMATICIAN WHO PAVED THE WAY FOR OTHER WOMEN TO EMPLOY THEIR TALENTS IN THESE FIELDS.'

NYC Dept of Environmental Protection Acting Commissioner Vincent Sapienza, 2017

Given the choice between a career in engineering and becoming a housewife, Nora Stanton Blatch Barney chose engineering. One of the first women to gain an engineering degree in the United States, she made a huge contribution to the water-supply system of New York City and was honoured for it in an unusual way.

Nora Stanton Blatch Barney was born on 30 September 1883 in Basingstoke, England. Barney's British father was the manager of a brewery. Her American mother and grandmother were both leading figures in the US suffragist movement, seeking the vote for women. Nora would follow in their footsteps and campaign for equal rights too.

In her teens, Barney went to Horace Mann School in New York, where she was an enthusiastic mathematics student, returning home to England for summer holidays. The family finally relocated to the States in 1902, as Barney enrolled at Cornell University, in Ithaca. Barney was one of the first women allowed into the Sibley School of Engineering. Three years later, she graduated with a degree in civil engineering with a highly praised thesis, 'An Experimental Study of the Flow of Sand and Water in Pipes under Pressure'.

Armed with her qualifications, Barney soon gained work above and below ground, designing bridges for the American Bridge Company, and subway

ABOVE: Nora Stanton Blatch Barney was the first American woman to receive a degree in civil engineering.

After further studies in mathematics and electricity at Columbia University, Barney assisted in the laboratory of Lee de Forest, the inventor of the radio vacuum tube. The pair married in 1908. Despite having an engineering degree, which her husband lacked, Nora was refused work in de Forest's capacitor factory. De Forest was not comfortable with his wife having a career. The couple separated a year later, with Nora giving birth to their daughter, Harriot shortly after.

Barney returned to engineering, as an assistant engineer and draughtsperson at the Radley Steel Construction Company for three years before acting as an architect, engineering inspector and structural-steel designer for the New York Public Service Commission in 1912.

In 1915, Barney took over as president of the Women's Political Union and campaigned and wrote in favour of equal rights for women, as editor of their publication *Women's Political World.* In 1919, Barney married for a second time, to Morgan Barnet, a naval architect. They relocated to Greenwich, Connecticut, where Nora Barney worked as a real-estate developer.

Despite her skills and qualifications, Barney often faced difficulty being appreciated as an equal by her colleagues. She was the first woman to be admitted to the American Society of Civil Engineers, but only as a junior member. It would not be until 1927 that they finally allowed full membership for women. (The ASCE posthumously awarded her fellow status in 2015.)

Barney died on 18 January 1971 in Greenwich. For her extensive contribution to New York City's water-supply infrastructure, she was honoured in 2017 by the city's Department of Environmental Protection who named their $30 million tunnel-boring machine 'Nora'. Barney's granddaughter Coline Jenkins said of the naming, 'Nora will be pushing forward and breaking ground, as she did in life.'

tunnels for a water supplier. At the American Bridge Company, Barney found herself the only woman in an office of 50 men, but she was well received and had ensured that she received the same wage as her male colleagues. Within three months she had taken charge of nine men and was visiting steel plants to monitor work. She was disappointed to find that both male and female workers were being exploited and underpaid in the factories and chose to move on. After nine months at the American Bridge Company, she passed the exam to work as assistant engineer at the Board of Water Works. Here, she acted as a drafting technician on plans for New York City's first reservoir and an aqueduct in the Catskill Mountains.

OLIVE DENNIS

'THERE IS NO REASON THAT A WOMAN CAN'T BE AN ENGINEER SIMPLY BECAUSE NO OTHER WOMAN HAS EVER BEEN ONE.'

Olive Dennis

GREATEST ACHIEVEMENTS

RECEIVED ENGINEERING DEGREE FROM CORNELL UNIVERSITY
She was only the second woman to receive an engineering degree from Cornell, and completed her course in just one year. 1920

JOINED B&O RAILROAD AS A DRAUGHTSPERSON IN THE ENGINEERING DEPARTMENT 1920

PROMOTED TO 'ENGINEER OF SERVICE' 1921

THE *CINCINNATIAN*
Dennis was responsible for overhauling the entire train to make it more amenable to passengers. 1947

ABOVE: Olive Dennis.

At the start of the 20th century, opportunities for women to make a career in engineering were very limited. Institutionalized barriers held women back from studying engineering in educational institutions and the industry remained male-dominated. Olive Dennis was just one of the determined women who overcame these obstacles to pursue a lifelong passion for engineering. She found her vocation in railroad engineering and went on to become an influential figure in transportation in the United States.

Born in Philadelphia in 1885, Olive Dennis's family moved to Baltimore when she was six. Her mother taught Olive needlework, but she was far more interested in making things. When Olive was given dolls to play with, she used her father's woodworking tools to build a dolls' house, complete with furniture. Her father then gave ten-year-old Olive her own set of tools for Christmas. She used them to build a working model streetcar for her younger brother. While on the way home from school, she sometimes stopped to watch the cranes and derricks at work on construction sites. These were the first signs of a growing fascination with engineering.

Dennis was a talented student. Following school she went to Goucher College in Baltimore, where she specialized in mathematics and science. She graduated at the top of her class. A scholarship then took Dennis to Columbia University in New York, where she studied mathematics and astronomy, gaining a master's degree. Now in her early twenties, she found a job teaching mathematics at McKinley Manual Training School in Washington DC. Dennis would be a teacher for almost a decade, but her childhood

BELOW: Dennis' engineering refinements transformed the performance and passenger experience on streamliner trains, such as the *Cincinnatian*.

dream of engineering had not been entirely forgotten. She attended summer schools during her vacations and studied surveying and civil engineering. When her younger brother mentioned that he was now studying engineering, Dennis decided it was time to finally follow her dream.

In 1919, Dennis quit her teaching job and enrolled to study structural engineering at Cornell University. It was a two-year degree course, but Dennis completed the course in just a year. At 35 years of age, she was finally a fully qualified engineer. When Olive walked up to receive her degree, a man in the audience scornfully asked, 'Now, what the heck can a woman do in engineering?' Dennis was quietly determined to show the world what a woman engineer could do. Her chance came when the Baltimore and Ohio Railroad gave her a job in their bridge-engineering section. It was a probationary role, but turned out to be the start of a career that spanned more than 30 years.

Dennis spent a year copying blueprints and designed her first railway bridge. Then Daniel Willard, the president of the Baltimore and Ohio Railroad offered her a new role. Railways were facing growing competition from buses and cars. Willard wanted Dennis to study the B&O Railroad and suggest improvements that would attract more female passengers. Dennis accepted the newly invented position of Engineer of Service and travelled the rail network for the next couple of years. In the first year alone she travelled more than 70,000 km (43,500 miles) on standard tickets to truly get a passenger's perspective. She made notes and recommendations on everything.

For some of the problems that Dennis identified, she found an engineering solution. To remedy stuffy carriages, she invented and patented the Dennis Ventilator, a type of window vent that gave passengers individual control over fresh air without obscuring the view from their carriage. Dennis was also instrumental in the B&O Railroad introducing the world's first air-conditioned trains in 1931. She improved passenger comfort on long-haul journeys by installing reclining seats and she made the layout of carriages more ergonomic. Her judgement was so highly valued that Dennis was asked to overhaul an entire train, the *Cincinnatian*. It went into service in 1947 with numerous improvements, including Dennis's design for an aerodynamic shroud, which transformed the old steam locomotive to a sleek modern streamliner.

Dennis retired in 1951, at the age of 65, having made her mark on railroad engineering. She died six years later. Throughout her long career, Dennis had been a vocal advocate for women in engineering. She blazed a trail that later female engineers went on to follow.

TACHŪ NAITŌ

GREATEST ACHIEVEMENTS

INDUSTRIAL BANK OF JAPAN
Head office, Tokyo 1923

NAITŌ HOUSE
Now the Tachū Naitō Memorial Museum 1926

NAGOYA TOWER
Now Japan's oldest TV tower 1954

SAPPORO TOWER
147 m (483 ft) 1957

TOKYO TOWER
Japan's tallest structure for 50 years 1958

ABOVE: Tachū Naitō.

'HIS PASSING MARKS THE END OF A CERTAIN ERA IN EARTHQUAKE ENGINEERING.'

George Housner, Fifth World Conference on Earthquake Engineering, 1970

'Father of Towers', Japanese architect and engineer Tachū Naitō designed rigid buildings and impressive steel towers that could withstand earthquakes. His ideas informed construction in earthquake-prone countries all around the world.

Tachū Naitō was born on 12 June 1886 in the village of Sakaki, Nakakoma, Japan. After high school he began studies in naval architecture at the Tokyo Imperial University (now University of Tokyo) before switching to conventional architecture after the Russo-Japanese War of 1904–05 all but wiped out the Japanese shipping industry.

Naitō started university just one year after a major earthquake had destroyed most of San Francisco, USA. Japan's position on the Pacific 'Ring of Fire' meant it too had to deal with regular earthquakes so Japanese architects had to design with this in mind. Naitō was determined to find a way to build earthquake-proof structures. His university dissertation was titled 'Theory of Earthquake Resistant Frame Construction'. While at university, Naitō was given a pocket-sized slide rule by his professor, the respected engineer Riki Sano. Naitō would continue to use this calculation aid throughout his career.

In 1912, Naitō became a professor of structural engineering at Waseda University, then continued his engineering studies by travelling the world to see how other countries dealt with earthquakes. He found few solutions but had his own breakthrough moment aboard a train. In 1917–18, crossing the States on the

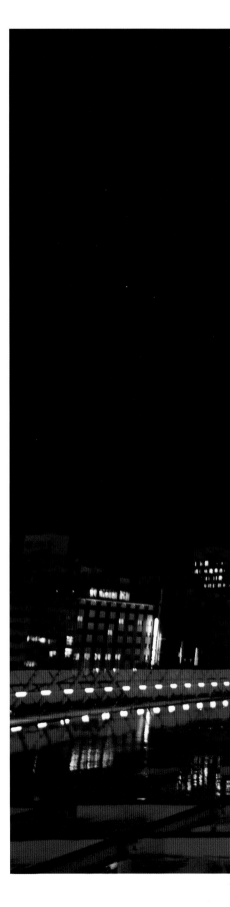

BELOW: Nagoya Tower, designed by Tachū Naitō and erected in 1954, is Japan's oldest TV tower.

First Transcontinental Railroad, Naitō observed how luggage was tossed around on the luggage racks when the train came to a sudden stop. To accommodate his books and paperwork, Naitō had removed the dividers from his trunk and seen it destroyed in transit. A second trunk, with dividers in place survived a later steamship journey intact. (The trunk, given to his son, became an exhibit in Naitō's house, now the Tachū Naitō Memorial Museum.) Naitō took the dividers idea and conceived of earthquake-proof shear walls within buildings, made of steel-reinforced concrete. By connecting these with beams, columns and flooring, he engineered rigid structures that held together even under the sideways forces exerted by earthquakes.

Naitō's methods were tested on the construction of a Kabuki theatre, Jitgugyo Building and the 30 m (98 ft)-tall head office of the Industrial Bank of Japan (designed by Setsu Watanabe) in 1923. Just three months after their completion, the Great Kantō earthquake occurred. At 7.9 magnitude, it was one of the most devastating earthquakes in Imperial Japan's history, causing widespread death and destruction. 700,000 buildings fell or were damaged yet, remarkably, all three of Naitō's buildings survived without damage.

Having been proved right, Naitō shared his theories of seismic

LEFT: The 147 m (483 ft)-tall Sapporo TV Tower was completed in 1957.

design with building designers in other earthquake-prone areas, such as California, USA. In 1926, he designed his own earthquake-proof three-storey home near Waseda University in Tokyo, with every floor constructed of reinforced concrete over a steel frame. But, it is for towers that Naitō's name is most known.

From 1925, and throughout his career, Naitō designed dozens of steel radio towers, with heights above 55 m (180 ft). These include what is now the oldest TV tower in Japan, the 180 m (590 ft)-tall Nagoya TV Tower (1954, famously torn down in two Godzilla movies). He also designed the 103 m (338 ft) Tsutenkaku 'Tower Reaching Heaven' in Osaka (1956) and the 147 m (483 ft)-tall Sapporo TV Tower (1957). His, literally, towering achievement is one of Tokyo's greatest landmarks. At 333 m (1,093 ft), the striking Tokyo Tower is taller than Paris's Eiffel Tower by 9 m (30 ft) and half its weight. Built of steel, including scrap taken from US tanks damaged in the 1950s Korean War, this radio and television transmitter remained the tallest structure in Japan until 2010.

In his later years, Naitō worked on plans for nuclear-power plants, including designs for the Calder Hall nuclear reactor in England, completed in 1956. He headed the Waseda University science and engineering division until his retirement in 1957. In 1964, Naitō was awarded a prestigious second-class Order of the Rising Sun.

Tachū Naitō died in Tokyo on 25 August 1970. Former students were among the many backers that funded a Naitō Memorial Hall at the university in his honour. But, perhaps, his greatest memorial is the number of buildings in Japan and other earthquake-prone countries around the world that have survived catastrophe due to Naitō's shared knowledge of seismic design.

VERENA
HOLMES

'SHE WAS ONLY NOW WINNING THE MERITS SHE HAD LONG SINCE EARNED BECAUSE THEY COULD NO LONGER BE WITHHELD.'

Claudia Parsons on Verena Holmes during World War II

GREATEST ACHIEVEMENTS

INSTITUTE OF MECHANICAL ENGINEERS
First female associate member 1924

WOMEN'S TECHNICAL SERVICE REGISTER
Helped provide female munitions workers 1942

HOLMES & LEATHER
Set up engineering firm for women 1946

ABOVE: Verena Holmes.

Keen on engineering from an early age, Verena Holmes found her opportunity to shine during two World Wars. Founding the Women's Engineering Society and her own business, she helped many women follow her lead into a career in draughting and engineering.

Verena Winifred Holmes was born on 23 June 1889, the daughter of a junior-schools inspector in Ashford, Kent, England. As a girl, she was said to have an interest in how things were made, taking apart her toy dolls, to understand how they were put together. Holmes studied at Oxford High School for Girls before taking work as a photographer. When World War I broke out, and young men were sent to fight on the Continent, women were needed for the jobs usually reserved for a male workforce. Keen on engineering, Holmes helped build wooden aircraft propellers at the Integral Propeller Company in Hendon. She also took night classes at the Shoreditch Technical Institute, East London.

Moving to Lincoln, Holmes continued her studies while working as a supervisor for 1,500 female employees at Ruston and Hornsby, a manufacturer of industrial engines. She persuaded the company to let her train in the fitting shops, gaining experience as a turner. After the war, many women returned to domestic duties as men returned from the front to take back their jobs. Holmes, however, was retained, and completed an apprenticeship as a draughtsperson by 1919. She also became a founding member of the Women's Engineering Society in this year. The Society was set up to help keep women in the workforce and to encourage them to take up engineering as a career. It proved particularly helpful two decades later, when women were needed in the factories when war broke out once more.

Three years later, Holmes graduated from Loughborough Engineering College with an engineering degree and was admitted to the Institute of Mechanical Engineers as an associate member in 1924, the first woman to receive the invitation. At university, Holmes became lifelong friends with fellow graduate Claudia Parsons, who received an automobile engineering degree. Verena and Claudia were two of just three women studying engineering at the university, alongside 300 men.

After graduating, for a time, Holmes worked for a marine-engineering company and as a technical journalist in the United States. She set up her own engineering consultancy business in 1925, with her specializations in marine and locomotive engineering. She patented several inventions including an artificial pneumo-thorax device for treating patients with tuberculosis, a surgeon's headlamp, an aspirator, a specialized gas valve for steam locomotives, and one

BELOW: Among Holmes' many patents was one for the Holmes-Wingfield pneumo-thorax apparatus, used to help patients suffering from the lung disease tuberculosis.

for internal combustion engines. In 1928, Holmes started working at the North British Locomotive Works in Glasgow, Scotland, followed by almost a decade at Research Engineers Ltd, three years later.

During World War II, Holmes took on major responsibility, working on rotary gyro valves for torpedoes, superchargers and other weapons technology for the British Navy. She also advised the government's Minister of Labour, Ernest Bevin, on training for female munitions workers. From 1940 to 1944 the Ministry of Labour appointed Holmes as their headquarters' technical officer, setting up the

ABOVE: Verena Holmes helped design the rotary gyro valves for torpedoes during World War II – an essential weapon for the British Navy in the Battle of the Atlantic.

RIGHT: After World War II, Holmes set up a company with Sheila Leather, providing employment for women in engineering, including this all-female factory in Gillingham, Kent.

Women's Technical Service Register in 1942.

After the war, in 1946, Holmes, and her colleague Sheila Leather, set up their own engineering business, Holmes and Leather. They employed only women in their metal-shearing factory in Gillingham, Kent. One of the company's many accomplishments was the design of the Safeguard Guillotine for use in cutting paper and card in schools.

As a manager and a writer, Holmes put much effort into encouraging women to take up training in engineering. She died on 20 February 1964. Five years later, the Women's Engineering Society, that she founded, used her legacy of £1,000 to launch an annual series of Verena Holmes lectures for young children, hoping to inspire a new generation of engineers. The yearly lectures continued for 40 years.

IGOR

SIKORSKY

'IF YOU ARE IN TROUBLE ANYWHERE IN THE WORLD, AN AIRPLANE CAN FLY OVER AND DROP FLOWERS, BUT A HELICOPTER CAN LAND AND SAVE YOUR LIFE.'

Igor Sikorsky

GREATEST ACHIEVEMENTS

S-2 SINGLE-ENGINE BIPLANE
First flight 1910

S-5 SINGLE-ENGINE BIPLANE
Gains pilot's licence 1911

THE GRAND
First successful four-engine aircraft 1913

S-29-A TWIN-ENGINE BIPLANE
First US design 1924

S-42 AMERICAN CLIPPER
Flying boat 1934

VOUGHT-SIKORSKY VS-300
Successful tethered helicopter flight 1939

R-4
World's first mass-produced helicopter 1942

ABOVE: Igor Sikorsky.

Russian-American aviation engineer and test pilot Igor Sikorsky's aircraft designs provided the template for the modern helicopter.

Born on 25 May 1889 in Kiev (now Kyiv, Ukraine), Igor Ivanovich Sikorsky had an early fascination with aviation. His father, a psychology professor, and mother, a qualified physician, encouraged Igor with his interests. Igor's mother shared her love of art and, in particular, the drawings of Leonardo da

BELOW: Sikorsky's first successful US aircraft, the S-29-A, was purchased by the record-breaking American aviator Roscoe Turner who used it to charter flights before converting it into a flying cigar store.

BELOW: Igor Sikorsky standing next to his prototype helicopter, the H-2, in 1910.

Vinci (see page 44). The Renaissance artist Da Vinci was also interested in engineering, and came up with elaborate plans for flying machines, including an early helicopter. Inspired, the 12-year-old Sikorsky built a small, rubber-band-powered helicopter that successfully took to the air.

Two years later, Sikorsky enrolled at the Saint Petersburg Naval Academy, but he left after three years, having decided to train as an engineer. In 1907, Sikorsky began studying at the Mechanical College of the Kiev Polytechnic Institute. Through the summer of 1908, he travelled across Europe with his father and learned about the aviation successes of the Wright brothers (see page 154) and Ferdinand von Zeppelin with his airships. Sikorsky was now determined to reach for the skies.

From 1909, Sikorsky started on the first of his helicopter designs, using a lightweight 25-horsepower engine and horizontal rotor. His attempt was not a success, which he blamed on a lack of available materials and funding. This project would have to wait a few more years to be developed. For now, Sikorsky turned his attention to fixed-wing aircraft, with a series of biplanes. On 3 June 1910, Sikorsky's S-2 biplane managed to take off for several feet. After some adjustments, Sikorsky was able to fly it to an altitude of 18–24 m (60–80 ft) before it stalled and crashed in a ravine. Not put off by any setbacks, Sikorsky continued to improve his designs. His S-5 single-engine biplane stayed airborne for over an hour. Now, Sikorsky also had enough experience to gain his pilot's licence.

After a mosquito blocked a fuel line in his S-6 model, forcing a crash-landing, Sikorsky chose to use more than one engine in future designs. His four-engine 'Grand' was a large biplane with a passenger compartment behind the cockpit. The Russian tsar, Nicholas II, came to watch the plane's inaugural flight, in 1913. As well

as being one of the first planes to include an enclosed cabin, the 'Grand' was the first successful four-engine aircraft, and the template for future commercial planes. A redesign of his four-engine 'Grand' was converted for use as a bomber in World War I.

After World War I, and following civil war and revolution in Russia, Sikorsky headed west for better opportunities in aviation development, arriving in New York, USA, on 30 March 1919. It would be several years before he would be able to return to aircraft design. In 1923, with the backing of several former Russian military officers, Sikorsky set up the Sikorsky Aero Engineering Corporation in Roosevelt, New York. A year later, with financial aid from the Russian composer Rachmaninoff, he produced the S-29-A, one of the first twin-engine aircraft to fly in the States. The

S-29-A (A for 'America') could carry 14 passengers and reach an air speed of 185 km/h (115 mph).

A 1926 attempt to design a prize-winning plane that could cross the Atlantic Ocean came to nothing, when the prototype veered off the runway and was engulfed in flames. Struggling to keep his business going, Sikorsky started work on amphibious aircraft. These proved a huge success, with major orders from Pan American Airways for their routes to Central and South America. Sikorsky built a new manufacturing plant in Stratford, Connecticut and agreed a deal with the United Aircraft and Transport Corporation.

Now a US citizen, Sikorsky developed the American Clipper in 1934, a large flying boat for

OPPOSITE: Russian-American aviation designer Igor Sikorsky insisted on piloting his prototype aircraft on their test flights.

ABOVE: The Sikorsky R-4 was the world's first mass-produced helicopter.

transatlantic flights, and returned to his obsession with helicopters. He filed several patents for his 'direct lift aircraft' between 1929 and 1935 before succeeding with a tethered test flight for his Vought-Sikorsky VS-300 on 14 September 1939. The VS-300 featured a single main rotor for lift and a smaller rotor attached to the tail to counteract torque (a rotating force), an arrangement that has been used in helicopters ever since. Sikorsky's triumph with the VS-300 led to the launch of the R-4 in 1942, the world's first mass-produced helicopter.

Sikorsky retired as an engineering manager in 1957, but continued acting as a consultant. He died on 26 October 1972, at his home in Easton, Connecticut, having fulfilled his dream of turning Leonardo da Vinci's sketches of a rotor-powered aircraft into reality. While he had much success with biplanes and flying boats, the Sikorsky name remains synonymous with the helicopter.

R. BUCKMINSTER
FULLER

GREATEST ACHIEVEMENTS

STOCKADE BUILDING SYSTEM 1927

DYMAXION HOUSE 1930

DYMAXION CAR 1933

FIRST GEODESIC DOME 1949

MONTREAL BIOSPHERE 1967

ABOVE: R. Buckminster Fuller.

'...ALL OF OUR BUILDINGS ARE ON A COMPRESSION BASIS. AND ALL OF THE ENGINEERING IS THAT WAY, AND THEY WILL NOT ACCREDIT TENSION. YET, I FOUND WHAT MAKES MY GEODESIC DOMES STAND UP *IS* THE TENSION...'

R. Buckminster Fuller, 'Everything I Know' lectures, 1975

Like earlier polymaths, the US engineer and inventor R. Buckminster Fuller was also something of a philosopher. He is chiefly remembered for the geodesic dome, reinventing an engineering structure that dates back to antiquity. But Fuller also wanted to engineer something much less tangible – the future. The concepts he learned from engineering fed into his wide-ranging philosophy and a humanist vision of technology and how it can shape our future. His ideas and theories are as much a part of his engineering legacy as the geodesic dome.

Born in Milton, Massachusetts in 1895, Buckminster Fuller came from a line of free-thinkers in an old New England family. Although super-smart and something of a boy-inventor, his nonconformist behaviour led to him being expelled from Harvard University. He took on various jobs, including working as a mechanic in a mill, and then served in the US Navy during World War I. Fuller started as a radio operator and went on to command a rescue boat, showing his flair as an inventor by designing a winch to recover downed pilots.

After his marriage to Anne Hewlett in 1917, Fuller went into business with his father-in-law, James Hewlett, who was an architect. Hewlett had designed a new modular system for building houses. They were assembled from hollow wooden blocks of compressed wood-shavings. Concrete was then poured into the hollows to fix the structure. Fuller built hundreds of houses with their Stockade Building System before the business failed in 1927. Facing financial ruin and still depressed following the death of his three-year-old daughter five years earlier, Fuller hit a low point and even contemplated suicide.

Fuller later claimed that he had a vision directing him to dedicate his life to humanity and credited this with helping him to bounce back. With this motivation Fuller started to look at the bigger picture of how technology could change the world to help people. He started by coming up with a new concept for an affordable factory-made house. Designed as a self-contained dwelling with its own water storage, natural ventilation and sanitation, he called his futuristic dwelling the Dymaxion House. It was to be built from aluminium with internal struts like a bicycle wheel, making it light enough to airlift into place.

The Dymaxion House was a radical concept, but, despite demonstrating promising prototypes, it didn't catch on. However, Fuller's idea of an autonomous house would remain influential and anticipated later eco-house designs. Another of Fuller's housing ideas, a prefabricated circular hut made of corrugated steel was more successful. Designed to ease a housing shortage during World War II, more than a hundred of his Dymaxion Deployment Units were commissioned and deployed by the US Army.

Fuller also designed a futuristic vehicle in the early 1930s – the super-streamlined three-wheel Dymaxion Car. It was also groundbreaking, but failed to progress beyond the prototype stage. Sadly, a fatal crash outside the Chicago World Fair in 1933 put an end to industry interest in developing Fuller's concept.

BELOW: The Dymaxion Car.

ABOVE: The Montreal Biosphere, one of Fuller's geodesic domes.

In 1949, Fuller built his first geodesic dome. The concept had already been developed by the German engineer Walther Bauersfeld and patented for a planetarium in 1922, but it was Fuller that realized its full potential. Traditional domes use compression to distribute their weight through sturdy supporting walls. The strength of these supports limits their size. A geodesic dome is essentially a hemisphere made from a lattice of triangular units. These spread the forces on the dome evenly throughout its structure, so it could be much larger.

A geodesic dome enclosed more space with an incredible economy of material compared to a traditional dome, essentially 'doing more with less'. This became a fundamental part of Fuller's philosophy, using the least energy and material to benefit the greatest number of people. He took out the US patent on the design and built a number of iconic domes, including the United States pavilion at the 1967 World Fair in Canada, which is the iconic Montreal Biosphere today.

Fuller planned houses using geodesic domes, but his dreams of futuristic cities, like many of his ideas, never really got further than the drawing board or development work. However, his geodesic domes and big ideas remain an inspiration for many.

IRMGARD
FLÜGGE-LOTZ

'I WANTED A LIFE WHICH WOULD NEVER BE BORING. THAT MEANT A LIFE IN WHICH ALWAYS NEW THINGS WOULD OCCUR.'

Irmgard Flügge-Lotz

GREATEST ACHIEVEMENTS

LOTZ METHOD
Wing lift calculation 1931

DISCONTINUOUS AUTOMATIC CONTROL
Publication of seminal work on on-off controls 1953

PROFESSOR OF ENGINEERING
First woman to take title at Stanford University, 1961

ABOVE: Irmgard Flügge-Lotz.

German-American aerodynamics expert and engineer Irmgard Flügge-Lotz provided calculations that helped in aircraft construction. She became the first female Professor of Engineering at Stanford University.

Born in Hamelin, Germany, on 16 July 1903, Irmgard Lotz showed an early talent for mathematics. The daughter of a journalist and an heir to a construction business, Irmgard was encouraged in her interest in technical subjects. When her father was drafted into the army during World War I, the young Lotz helped support her family by working as a maths and Latin tutor while studying at the Hannover Gymnasium. In 1923, she enrolled at the Leibniz University in Hannover as a mathematics and engineering student, staying on to earn her doctorate in 1929. She was often the only woman in her class.

Lotz started work at the aeronautical research company Aerodynamische Versuchsanstalt (AVA) in Göttingen as a junior research engineer, working closely with the directors Ludwig Prandtl and Albert Betz. She soon proved her worth, delivering the solution to a mathematical problem that had dogged her older colleagues for years. Lotz's calculation made it easier for the engineers to work out the distribution of lift over an aeroplane's wing. The calculation

became known as the 'Lotz method' and led to Lotz being promoted to unofficial head of the department, responsible for choosing further research programmes.

In 1938, Lotz married Wilhelm Flügge, a civil engineer. With Adolf Hitler's Nazi Party in power in Germany, Flügge struggled to gain promotion at work because of his anti-Nazi views. Having become Head of the Department of Theoretical Aerodynamics, Irmgard Flügge-Lotz also found her progress frustrated. She was denied the position of research professor purely because of her sex. The Flügge-Lotzes decided to leave Göttingen and move to the German capital Berlin where Irmgard became a consultant in aerodynamics at the Deutsche Versuchsanstalt für Luftfahrt (DVL). War broke out soon after their move. By 1944, Allied bombing raids on Berlin forced the Flügge-Lotzes and their department to move to Saulgau in the southern German countryside.

After the war Saulgau came under French control and both Flügge-Lotz and her husband were invited to help France's aeronautical research programme. They moved to Paris in 1947 to join the Office National d'Etudes et de Recherches Aerospatiales (ONERA). Flügge-Lotz acted as chief of one research group, working on automatic control theory. A year later the Flügge-Lotzes moved to the United States, where both lectured at Stanford University. An antiquated rule at the university about married couples would not allow Flügge-Lotz to become a professor, as her husband did.

Despite this lack of official title, at Stanford, Flügge-Lotz led research projects and ran seminars, guiding students with their dissertations in aerodynamic theory. She taught her first Stanford course in 1949

BELOW: A mathematical genius and expert in the field of aerodynamics, Irmgard Flügge-Lotz became Stanford University's first female professor in engineering.

and provided courses in mathematical hydro- and aerodynamics for graduate students. Flügge-Lotz continued to show strong interest in fluid mechanics, numerical methods and automatic controls. Her work also involved the use of computers. While continuing to investigate her chosen field, Flügge-Lotz was a motivational figure on campus, inviting students to her house for regular informal study groups.

In 1960 Flügge-Lotz was the only female delegate from the United States at the first Congress of the International Federation of Automatic Control in Moscow, a 'lecturer' among a group of professors. This discrepancy was finally corrected for the next term, when she became Stanford's first female Professor of Engineering.

Flügge-Lotz retired in 1968 but continued as a researcher, studying satellite control systems, heat transfer and drag in high-speed vehicles. In 1970, she was elected a Fellow of the American Institute of Aeronautics and Astronautics (AIAA). Flügge-Lotz suffered from increasing arthritis after her retirement. She died in Palo Alto, California on 22 May 1974 after a long illness. Over her career she published over 50 technical papers and wrote two books. Her persistence and mathematical expertise had resulted in great progress in the field of aerodynamics. Forty years after her death, Stanford University paid tribute to Irmgard Flügge-Lotz, describing her as one of 35 'Engineering Heroes' who had delivered great progress in technology and science.

FRANK WHITTLE

GREATEST ACHIEVEMENTS

FUTURE DEVELOPMENT OF AIRCRAFT DESIGN
Whittle's thesis on the theory behind the jet engine was written when he was just 21 years old. 1928

WHITTLE'S PATENT FOR JET ENGINE 1930

FIRST TEST OF WHITTLE'S JET PROPULSION ENGINE
1937

HEINKEL HE 178
The first jet-powered flight was made by an aircraft using Von Ohain's turbojet engine. 1939

GLOUCESTER E28/39 AIRCRAFT
An aircraft powered by Whittle's W1 turbojet engine. 1941

GLOUCESTER METEOR JETS ENTER SERVICE 1944

ABOVE: Frank Whittle.

'I'D COME TO THE CONCLUSION THAT IF YOU WANTED TO GO BOTH FAST AND FAR, YOU'D NEED TO GO VERY HIGH, TO HEIGHTS WHERE THE PISTON ENGINE WOULDN'T WORK AND AT SPEEDS WHERE THE PROPELLER WOULD BE VERY INEFFICIENT.'

Sir Frank Whittle, 1986

Every jet aeroplane that crosses the sky today, flies thanks to the pioneering work of two aviation engineers: Sir Frank Whittle and Hans von Ohain. Working independently in Britain and Germany respectively, not least because they were on opposing sides during World War II, these two engineers made a huge leap forward by developing the jet engine.

British engineer Frank Whittle was the first to come up with the idea of using a jet propulsion gas turbine engine for aeroplanes. He patented his design in 1930 and successfully tested an engine on the ground on 12 April 1937. Sadly, Whittle's ambitions to develop an aeroplane powered by his turbojet engine were frustrated and delayed by the British Air Ministry. This meant that Von Ohain was the first to actually build and fly a jet-powered aeroplane. His turbojet engine powered the Heinkel He 178 aircraft that made the world's first turbojet-powered flight on 27 August 1939.

Whittle's achievement is due to his tenacity as well

BELOW: Whittle's jet propulsion engine.

as his engineering talents. Born into a working-class family in Coventry, he had a passion for flight from an early age and built model aeroplanes. As a boy, he picked up some knowledge of engineering by helping his father, Moses, a foreman at a machine tool factory who later owned a small engineering business. Frank was smart, but he disliked school and homework. Only flying fired his imagination. He devoured books on the subject and taught himself about turbines, engine mechanics and the theory of flight in the public reference library.

As soon as he had finished school, Whittle was determined to become a pilot. He applied for an apprenticeship at the RAF College at Cranwell, but was rejected as he didn't meet the physical criteria for height and stature. He followed the advice of a sympathetic RAF physical-training instructor and adhered to a strict diet and exercise programme to help him bulk up. Whittle later claimed that it put 3 inches on his height. However, the RAF still rejected him, saying it couldn't accept his application a second time. Undeterred, Whittle applied a third time, but

LEFT: The Gloucester E.28/39 was the first British jet aircraft and was powered by Whittle's revolutionary turbojet engine.

with a slightly different name. This time he was finally accepted.

It was while he was an apprentice at Cranwell that Whittle started to consider the possibilities of gas turbines as a power plant for aeroplanes. In 1928, he laid out the theoretical groundwork for a turbojet in a thesis he wrote, entitled *Future Developments in Aircraft Design*. He was 21 years old.

As a pilot, Whittle knew very well the limits of conventional piston-engine aeroplanes, which were reaching the limits of their performance. Whittle's breakthrough was to realize that at a high altitude, with reduced air resistance and at speeds higher than a propeller could handle, a turbine engine could perform well. The principles of jet propulsion and the gas turbine engine were already well known, but Whittle combined them in a new way to solve a new problem.

Whittle's design for a turbojet engine had a fan-like compressor at the front. This sucked air into a ring of combustion chambers, where it was compressed with fuel and burned, producing exhaust gases that created thrust and also drove a turbine. The turbine was mounted on the same central shaft as the compressor, so it also powered it to draw in air.

Frustratingly, Whittle's turbojet idea found little support when he tried to promote it within the Air Ministry and the wider aviation industry. The accepted idea was that gas turbines were impractical, and that the materials required to build an engine to withstand the stresses and temperatures of jet propulsion didn't exist yet. Whittle struggled on with

a limited amount of finance and support. Meanwhile, Hans von Ohain had found more enthusiasm for his experimental jet engine in Germany. He teamed up with the aircraft manufacturer Ernst Heinkel and as a result was able to see his turbojet engine power the first flight in 1939.

Whittle was finally able to get a Gloucester E28/39 aircraft powered by his W1 turbojet engine into the air on 15 May 1941 for a 17-minute test flight, some two years later. Unlike von Ohain's experimental engine, Whittle's W1 was a reliable, production engine. It went on to be fitted to Gloucester Meteor jets, which went into service in 1944. After the war, von Ohain said that the Battle of Britain might not have happened if the Air Ministry had listened to Whittle as Britain would have clearly had air superiority with a jet fighter.

In 1935, after his turbojet engine design had been rejected by the British Air Ministry, Whittle formed a company called Power Jets to try and develop jet aeroplanes. He could see their potential and was already envisaging a future where passenger jets were crossing the Atlantic. By this time, he was also studying engineering at Cambridge University. As if that wasn't enough work and pressure, he had also to contend with trying to finance Power Jets, as well as being their chief engineer. Difficulties with the Air Ministry continued to stifle development and a deal it made with the engine manufacturer Rover led to further delays and problems. At one stage Whittle's original patent lapsed and he hadn't sufficient money to renew it. Rolls-Royce eventually took on the further development of Whittle's engine in 1943, but in order to further his idea of a British jet, Whittle had to generously relinquish his interests in his company when it was nationalized. Although he was awarded £100,000 in 1948 for his groundbreaking work by a Royal Commission on Awards to Inventors, Whittle never collected any royalties for his invention.

Ultimately, Whittle's vision of a jet-age future was proved true. His jet engine went on to revolutionize the aircraft industry, just as he had foreseen as a junior flying officer. Today, the impact of Whittle's invention is visible in the skies above us – and in our global economy. We live in a world that has been shaped by fast and affordable air transport.

WERNHER VON BRAUN

'IT WILL FREE MAN FROM HIS REMAINING CHAINS, THE CHAINS OF GRAVITY WHICH STILL TIE HIM TO THIS PLANET. IT WILL OPEN TO HIM THE GATES OF HEAVEN.'

Wernher von Braun on the rocket

GREATEST ACHIEVEMENTS

V-2 ROCKET
Ballistic missile 1942

REDSTONE
First ballistic missile for the US 1953

EXPLORER 1
Redstone launches first US satellite 1958

NASA
Von Braun becomes director for US space agency 1960

SATURN V
Launch of multi-stage heavy-lift rocket 1967

APOLLO 11
Saturn V rocket launches first moon landing mission 1969

ABOVE: Wernher von Braun.

One of the most-significant figures in early rocket development, Wernher von Braun helped the US space programme achieve its goals of sending a man to the moon but his reputation is tarnished by his earlier work in Germany.

Born into German nobility on 23 March 1912 in Wirsitz, Germany (now Wyrzysk, Poland), Wernher von Braun began looking to the stars from an early age. After a move to Berlin, Wernher's mother gave him a telescope as a gift on his 13th birthday and the young von Braun became fascinated with astronomy. After reading *The Rocket into Planetary Space* by the Austro-Hungarian engineer Hermann Oberth, von Braun put his mind to serious study of physics and mathematics, in the hope of becoming a rocket engineer.

In 1930, von Braun got to work with Oberth, when he joined the Spaceflight Society at the Berlin Institute of Technology, assisting in liquid-fuelled rocket tests. Two years later, von Braun graduated with a diploma in mechanical engineering, then moved on to the Friedrich-Wilhelm University of Berlin for further physics and engineering studies. While here, von Braun's

BELOW: German-American rocket designer Wernher von Braun helped lead the US space programme following World War II.

BELOW: The V-2 rocket, designed by von Braun and Klaus Riedel, was used as a ballistic missile during World War II. From September 1944 to March 1945, thousands were launched at London, Antwerp and Liege, killing an estimated 9,000 people. Twelve thousand forced labourers died during their manufacture.

BELOW: The V-2 rocket, designed by von Braun and Klaus Riedel, was used as a ballistic missile during World War II. From September 1944 to March 1945, thousands were launched at London, Antwerp and Liege, killing an estimated 9,000 people. Twelve thousand forced labourers died during their manufacture.

amateur rocketing group caught the eye of army ordnance officer Walter Dornberger, who offered the student a research grant and the chance to test at an army ground south of Berlin. By 1935, von Braun's group had managed to fire two test rockets above 2.4 km (1.5 miles) using liquid fuel. The research was included in von Braun's degree dissertation, which was classified until 1960 due to its sensitive military contents.

In 1933, Adolf Hitler came to power in Germany. To be allowed to continue his work with rockets, von Braun joined the Nazi Party in 1937 and relocated to a secret military facility in Peenemünde on the Baltic Sea, with Dornberger as military commander and von Braun as technical director.

In 1939, World War II broke out. Von Braun's work was now steered towards military applications.

Following successful tests on von Braun's A-4 long-range ballistic missile, the Nazi Propaganda Ministry renamed it Vergeltungswaffe-Zwei, Vengeance Weapon 2. The V-2 flew at over 5,600 km/h (3,500 mph) and could deliver a 980 kg (2,200 lb) warhead to a target 320 km (200 miles) away. Twelve thousand prisoners from a nearby concentration camp were employed and died in an underground factory in Mittelwerk to build the V-2. Around 2,800 V-2 rockets were fired upon Belgium and Britain from 1944, resulting in much destruction and an estimated 9,000 deaths. How much von Braun knew of factory conditions, and how responsible he felt for the deaths caused by his missiles is unclear. Certainly, any regard for von Braun's achievements in rocket science is tempered by knowledge of his aid to the Nazi cause.

At the end of the war, von Braun surrendered to

the Americans and spent the next 15 years working for the US Army on their ballistic missile programme, developing the Redstone, Jupiter-C, Juno and Pershing missiles. In 1952, he wrote the first of a series of articles about rockets and space travel for *Collier's Weekly* magazine. In these von Braun described orbiting space stations and moon bases. He also published a book envisioning a manned mission to Mars. In 1955, now a US citizen, von Braun's spacefaring ideas were brought to life through animation when he introduced an edition of Disney's *Tomorrowland* TV series.

Throughout the '50s and '60s, the United States was involved in a 'Space Race' with the Soviet Union and happy to employ von Braun to gain an advantage in rocket technology. The Soviets were first to put a satellite, *Sputnik I*, into Earth orbit in 1957. Von Braun's Redstone rocket launched the first US satellite, *Explorer 1*, a year later. The Russians then put a man into space (Yuri Gagarin, 1961) three weeks ahead of the US. The next target was the moon.

On 16 July 1969, a Saturn V rocket, designed by von Braun and his team, launched a trio of astronauts into space. Four days later, two of them – Neil Armstrong and Edwin 'Buzz' Aldrin – landed safely on the moon's surface. Von Braun's childhood dreams of reaching other worlds had come to be. After five further missions had delivered astronauts to the moon and returned them safely to Earth, the Apollo programme was cancelled. Disappointed, von Braun retired from NASA in 1972, and took on the post of Vice President for Engineering and Development at the aerospace company Fairchild Industries in Maryland.

In 1977, von Braun was awarded the National Medal of Science in Engineering by President Ford but was too ill to attend the White House ceremony. Wernher von Braun died of pancreatic cancer in Alexandria, Virginia, on 16 June 1977. While his association with the German Nazi Party cannot be overlooked, his forward-thinking development of rocket technology had helped take humankind to the moon.

OPPOSITE: Wernher von Braun stands by the F-1 engines of the Saturn V rocket.

RIGHT: The three-stage Saturn V super heavy-lift launch vehicle, designed by von Braun and his team at NASA, was used between 1967 and 1973, helping to take 24 astronauts to the moon.

FAZLUR RAHMAN KHAN

'RARELY HAS ANY ENGINEER PLAYED AS KEY A ROLE IN THE SHAPING OF ARCHITECTS' IDEAS AND THE SHAPING OF BUILDINGS THEMSELVES.'

American Institute of Architects, posthumous award nomination, 1983

Fazlur Rahman Khan was a structural engineer, whose innovative ideas revolutionized the construction of tall buildings. Born in Bangladesh, he emigrated to the US, where in the 1960s he developed a fundamentally new way to build skyscrapers. Khan pioneered the use of tubular structures instead of a traditional framework of steel boxes. This significantly reduced the steelwork required and meant his skyscrapers could be both taller and more economical to build. Khan's visionary breakthrough started a new era of high-rise construction. Most tall buildings constructed today still rely on the principles he established.

Khan was born in 1929, in a region of India that would become modern-day Bangladesh. His father was a respected teacher of mathematics and helped to nurture the young Khan's talents, turning the chore of homework into something more challenging and entertaining. As a boy, Khan excelled at mathematics and physics. He also enjoyed tinkering with anything mechanical. Encouraged by his father, Khan later chose to study civil engineering rather than physics. He was the top student in his class when he graduated in 1950.

Two years of teaching followed, while Khan also gained practical experience as an assistant

LEFT: The John Hancock Tower in Boston.

OPPOSITE: The Sears Tower.

engineer in highway construction. In 1952, Khan was awarded two scholarships that enabled him to travel to the US where he studied for three years at the University of Illinois. He took two master's degrees: one in theoretical and applied mechanics, the other in structural engineering. He also gained a Ph.D in

structural engineering. Khan's intense studies in both theoretical and practical engineering gave him a solid grounding that helped him to look beyond conventional solutions as a structural engineer.

While Khan was considering job offers from some of the most prestigious engineering firms in the US,

LEFT: The Hajj terminal of King Abdulaziz airport in Saudi Arabia.

their offices, pitched for a job and was so impressive they made an offer on the spot. Khan accepted, even though the salary was lower than other offers. He was swayed by the chance to dive in at the deep end and have responsibility for his own projects right away. Khan started in 1955 and worked there for a year and a half. A brief spell as an executive engineer in Pakistan followed, before he re-joined Skidmore, Owings and Merrill for good in 1960.

Back in Chicago, Khan started what would be a long creative partnership with the architect Bruce Graham. It was while working closely with him that Khan made the breakthrough that rewrote the rulebook for engineering tall buildings. When Graham asked what would be the most economical design possible for a skyscraper, Khan responded that it would be like a tube. Khan instantly grasped the advantages of tubular design. By making a skyscraper's outer wall strong, there was no longer a need for lots of interior columns and steelwork. Tubular skyscrapers needed less material, their structure freed up interior floorspace and made a building that better stood up to the forces of the wind. One story goes that Khan may have been partly inspired by the strength of the tall, cylindrical bamboo stems he had seen near his home as a boy. But insight was also founded on deep theoretical and practical knowledge.

The first steel tubular design skyscraper he built was the 100-storey John Hancock Building in Chicago, completed in 1969. Chicago's Willis Tower (formerly the Sears Tower) followed in 1973, which used a modular tubular design with nine rectangular tubes bundled together. It was the world's tallest building for 25 years. Khan devised other tubular-design variants and used steel cross-bracings to reinforce some of his tall buildings.

Khan built other iconic structures before his untimely death in 1982, at the age of 52, including the famous Hajj terminal of King Abdulaziz International Airport in Saudi Arabia. But it is for his work with skyscrapers that Khan is chiefly remembered and for which he is acclaimed as one of the greatest structural engineers in history.

he chanced to bump into a friend who was working at Skidmore, Owings and Merrill, an architectural firm based in Chicago. When Khan heard about the projects the firm were involved with and how they integrated architecture and structural engineering, he knew he wanted to be involved. He showed up at

PICTURE CREDITS

t = top, b = bottom, l = left, r = right

AKG Images: 23

Alamy: 14, 17, 26b, 36, 42, 48, 51, 63, 72, 76, 78, 113, 116, 151, 163, 164b, 183, 188

Bridgeman Images: 49, 64

British Library: 57b

Brooklyn Museum: 99

Flickr: 58

Getty Images: 13, 16t, 19, 25t, 25b, 37, 41, 43, 53, 59, 60l, 62, 69, 70, 80, 83t, 84, 86, 93, 94, 102, 108b, 110, 124, 127, 147, 158, 165, 176, 182, 196, 197

Imperial War Museum: 135

Library of Congress: 83b, 85, 95, 97, 111, 118, 120, 123, 125, 134, 144, 154t, 155, 160, 172, 173, 191

Metropolitan Museum of Art, New York: 8

NASA: 166b, 167, 169, 170, 200b, 202, 203

Public Domain: 162, 194

Science and Society Picture Library: 121

Science Photo Library: 201l, 190

Smithsonian Institution: 157 (National Air & Space Museum)

Shutterstock: 2t, 2bl, 6, 7, 10, 11bl, 22, 30, 35, 46, 53b, 73, 74, 88, 101, 104, 108t, 109, 132, 177, 179, 181, 195, 205, 206

US Patent Office: 141, 143, 144

Wellcome Collection: 87, 90t, 90b, 128b, 139

Wikimedia Commons: 2br, 9, 11t, 11br, 12, 16b, 18t, 18b, 20r, 21, 24, 26t, 27, 28, 29, 31, 32, 33, 34, 38, 40, 44, 45, 47t, 47b, 50t, 50b, 52t, 52b, 53t, 54, 56, 57t, 60r, 61t, 61b, 65, 66, 67, 68, 71, 77, 81, 82, 91, 92, 98t, 98b, 100, 103, 105, 107, 112, 114, 115, 117, 119, 122, 128t, 129l, 129r, 130, 131, 133, 136t, 136b, 137, 140, 142t, 146, 148, 149, 150, 152, 153, 154b, 156, 159, 161, 164t, 166t, 168, 171, 174, 175, 178, 180, 184t, 184b, 185, 186, 189, 192, 193, 198, 200t, 201, 204